Phillip I. Good

Resampling Methods
A Practical Guide to Data Analysis

Birkhäuser
Boston • Basel • Berlin

Phillip I. Good
205 West Utica
Huntington Beach, CA 92648
USA
brother_unknown@yahoo.com

Library of Congress Cataloging-in-Publication Data

Good, Phillip I.
    Resampling methods : a practical guide to data analysis / Phillip
I. Good.
       p. cm.
    Includes bibliographical references and index.
    ISBN 0-8176-4091-6 (hardcover : alk. paper)
    1. Resampling (Statistics)   I. Title.
QA278.8.G66   1999
519.5—dc21   98-26978
                                             CIP

AMS Subject Classifications: 62G

Printed on acid-free paper.

© 1999 Birkhäuser Boston   ***Birkhäuser***

ISBN 0-8176-4091-6
ISBN 3-7643-4091-6

Formatted from the author's EXP files.
Printed and bound by Maple-Vail Book Manufacturing Group, York, PA.
Printed in the United States of America.

9 8 7 6 5 4 3 2 1

Contents

Preface

Intended for class use or self-study, this text aspires to introduce statistical methodology—estimation, hypothesis testing, and classification—to a wide audience, simply and intuitively, through resampling from the data at hand.

The resampling methods—permutations, cross-validation, and the bootstrap—are easy to learn and easy to apply. They require no mathematics beyond introductory high-school algebra, yet are applicable in an exceptionally broad range of subject areas.

Introduced in the 1930's, their numerous, albeit straightforward and simple calculations were beyond the capabilities of the primitive calculators then in use; they were soon displaced by less powerful, less accurate approximations that made use of tables. Today, with a powerful computer on every desktop, resampling methods have resumed their dominant role and table lookup is an anacronism.

Physicians and physicians in training, nurses and nursing students, business persons, business majors, research workers and students in the biological and social sciences will find here a practical guide to descriptive statistics, classification, estimation, and testing hypotheses.

For advanced students in biology, dentistry, medicine, psychology, sociology, and public health, this text can provide a first course in statistics and quantitative reasoning;

For industrial statisticians, statistical consultants, and research workers, this text provides an introduction and day-to-day guide to the power, simplicity, and versatility of the bootstrap, cross-validation, and permutation tests.

Hopefully, all readers will find my objectives are the same as theirs:

To use quantitative methods to characterize, review, report on, test, estimate, and classify findings.

If you're just starting to use statistics in your work, begin by reading chapters 1–5 which cover descriptive statistics, cause and effect, sampling, hypothesis testing and estimation, portions of chapter 6 with its coverage of the essential, but challenging ideas of significance level, sample size and power, chapter 10 on classification, and chapters 7 on contingency tables and/or 8 on experimental design depending upon your interests. Recurrent themes—for example, the hospital data considered in the exercises for chapters 1–2 and 5–8, tie the material together and provide a framework for self-study and classes.

For a one-quarter short course, I took the students through chapter 1—we looked at, but did not work though section 1.2 on charts and graphs, chapter 2, letting the students come up with their own examples and illustrations, chapter 3 on hypothesis testing (3.1–3.3 and 3.5–3.7), chapter 4 (reviewing section 2.4), and sections 5.1, 5.2, and 5.5 on bootstrap estimation. One group wanted to talk about sample size (section 6.2–6.4), the next about contingency tables (section 7.1, 7.3, 7.5).

Research workers, familiar with the material in chapters 1 and 2, should read chapters 3, 5 and 6, and then any and all of chapters 7–11 according to their needs. If you have data in hand, turn first to chapter 12 whose expert system will guide you to the appropriate sections of the text.

A hundred or more exercises included at the end of each chapter plus dozens of thought-provoking questions will serve the needs of both classroom and self-study. C++ algorithms, Stata, SPlus, SC and SAS code and a guide to off-the-shelf resampling software are included as appendices. The reader is invited to download a self-standing IBM-PC program which will perform most of the permutation tests and simple bootstraps described here. The software is self-guiding, so if you aren't sure what method to use, let the program focus and limit your selection. To obtain your copy of the software, follow the instructions on my home page http\\users.oco.net\drphilgood. To spare you and your students the effort of retyping, I've included some of the larger data sets, notably the hospital data (section 1.8) and the birth data (section 10.9), along with the package.

My thanks to Symantek, Cytel Software, Stata, and Salford Systems without whose GrandView™ outliner, StatXact™ Stata™, and CART™ statistics packages this text would not have been possible.

I am deeply indebted to Bill Sribney and Jim Thompson for their contributions, to John Kimmel and Lloyd S. Nelson for reading and commenting on portions of this compuscript, and to the many readers of *Permutation Tests,* my first text, who encouraged me to reach out to a wider audience.

Phillip I. Good
Huntington Beach, Fullerton, San Diego, San Ramon, and Torrance, California

CHAPTER 1

Descriptive Statistics

1.0. Statistics

Statistics help you

- decide what data and how much data to collect.
- analyze your data.
- determine the degree to which you may rely on your findings.

A time-worn business adage is to begin with your reports and work back to the data you need. In this chapter, you'll learn to report your results through graphs, charts, and summary statistics and to use a sample to describe and estimate the characteristics of the population from which the sample is drawn.

1.1. Reporting Your Results

Imagine you are in the sixth grade and you have just completed measuring the heights of all your classmates.

Once the pandemonium has subsided,[1] your instructor asks you and your team to prepare a report summarizing your results.

Actually, you have two sets of results. The first set consists of the measurements you made of you and your team members, reported in centimeters, 148.5, 150.0, and 153.0. (Kelly is the shortest incidentally, while you are the tallest.) The instructor

---

[1] I spent the fall of 1994 as a mathematics and science instructor at St. John's Episcopal School in Rancho Santa Marguarite, CA. The results reported here, especially the pandemonium, were obtained by my sixth-grade homeroom. We solved the problem of a metric tape measure by building our own from string and a meter stick.

TABLE 1.1. Heights of Students in Dr. Good's Sixth-Grade Class in cm.

141, 156.5, 162, 159, 157, 143.5, 154, 158, 140, 142, 150, 148.5, 138.5, 161, 153, 145, 147, 158.5, 160.5, 167.5, 155, 137

Summarize: To describe in a straightforward, easy-to-comprehend fashion.

---

asks you to report the *minimum*, the *median*, and the *maximum* height in your group. This part is easy, or at least it's easy once you look up *median* in the glossary of your textbook and discover it means "the one in the middle." In your group, the minimum height is 148.5 centimeters, the median is 150.0 centimeters, and the maximum is 153.0 centimeters. *Minimum* means smallest, *maximum* means largest, *median* is the one in the middle. Conscientiously, you write these definitions down–they could be on a test.[2]

Your second assignment is more challenging. The results from all your classmates have been written on the blackboard—all 22 of them (see Table 1.1). Summarize these results, your teacher says.

You copy the figures neatly into your notebook. You brainstorm with your teammates. Nothing. Then John speaks up—he's always interrupting in class. Shouldn't we put the heights in order from smallest to largest? "Of course," says the teacher, "you should always begin by ordering your observations." This would be an excellent exercise for the reader as well, but if you are impatient, take a look at Table 1.

"I know what the minimum is," you say (come to think of it, you are always blurting out in class, too), "137 millimeters, that's Tony."

"The maximum, 167.5, that's Pedro: he's tall," hollers someone from the back of the room. As for the median height, the one in the middle is just 153 centimeters (or is it 154)?

1.2.   Picturing Data

My students at St. John's weren't finished with their assignments. It was important for them to build on and review what they'd learned in the fifth grade, so I had

TABLE 1.2. Heights of Students in Dr. Good's Sixth-Grade Class in cm, Ordered from Shortest to Tallest.

137.0 138.5 140.0 141.0 142.0 143.5 145.0 147.0 148.5 150.0 153.0 154.0 155.0 156.5 157.0 158.0 158.5 159.0 160.5 161.0 162.0 167.5

---

[2] A hint for self-study, especially for the professional whose college years are long behind him or her: Write down and maintain a list of any new terms you come across in this text along with an example of each term's use.

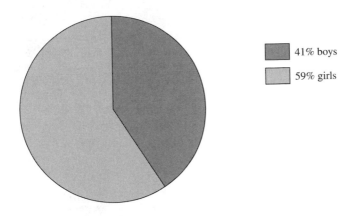

FIGURE 1.1. Students in Dr. Good's Sixth-Grade Class. a) Pie Chart—distribution by sex. The area of each shaded area corresponds to the relative frequency of that group in the sample.

them draw pictures of their data. Not only is drawing a picture fun, but pictures and graphs are an essential first step toward recognizing patterns. In the next few sections and again in Chapter 2, we'll consider a variety of different ways you can visualize your data.

1.2.1. Graphs

You're probably already familiar with the pie chart, bar chart, and cumulative frequency distribution. Figure 1.1.a, b, d, and e, prepared with the aid of Stata[TM], are examples.[3] The box and whiskers plot, Figure 1.1.c, and the scatter plot, Figure 1.2, may be new to you.

The pie chart illustrates percentages; the area of each piece corresponds to the relative frequency of that group in the sample. In Figure 1.1.a, we see that 41% of my class were boys and the remaining 59% were girls.

The bar chart (Figure 1.1.b) illustrates frequencies; the height of each bar corresponds to the frequency or number in each group. Common applications of the bar chart include comparing the GNPs of various countries or the monthly sales figures of competing salespersons.

---

[3] Details of how to obtain copies of Stata and other statistics packages cited in this text may be found in Appendix 1.

FIGURE 1.1. b) Bar Chart—distribution by sex. The height of each bar corresponds to the frequency of that group in the population.

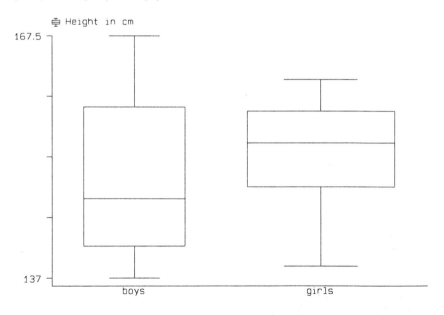

FIGURE 1.1. c) Box and Whiskers Plot of Student Heights—by sex. Each box extends between the 25th and 75th percentile of the corresponding group. The line in the center of each box points to the median. The ends of the whiskers highlight the minimum and the maximum. The width of each box corresponds to the frequency of that group in the population.

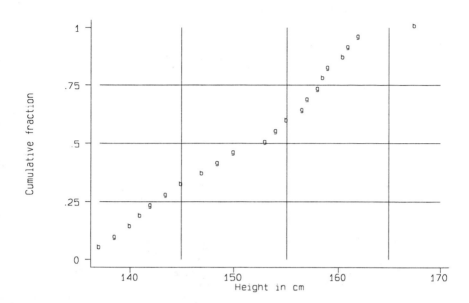

FIGURE 1.1. d) Cumulative Frequency—distribution of heights. The symbols *b* and *g* denote a boy and a girl, respectively.

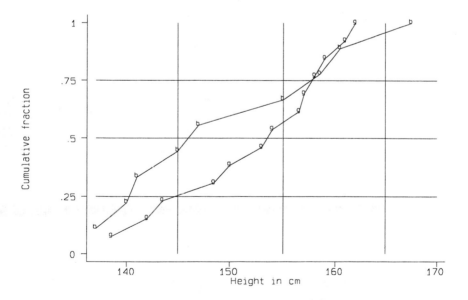

FIGURE 1.1. e) Cumulative Frequency—distributions of student heights by sex.

The box and whiskers plot (Figure 1.1.c) provides the minimum, median, and maximum of each group side by side in chart form for comparison. The box extends over the *interquartile range* from the 25th to the 75th percentile. The line in the middle is the median. The whiskers reach out to the minimum and maximum of each sample. This particular plot allows us to make a side-by-side comparison of the heights of the two groups that make up my classroom. (Note that I had more girls than boys in my class and that in this age group, the girls tend to be taller, although both the shortest and the tallest students are boys.)

The cumulative frequency distribution (Figure 1.1.d, e) starts at 0 or 0%, goes up one notch for each individual counted, and ends at 1 or 100%. A frequent application, part of the modeling process, is to compare a theoretical distribution with the distribution that was actually observed.

Sixth graders make their charts by hand. You and I can let our computers do the work, either via a spreadsheet like Excel™ or Quarto™ or with a statistics program like Stata™ or SPSS™.

Figure 1.2, a scatter plot, allows us to display the values of two observations on the same person simultaneously. Again, note how the girls in the sixth grade form a taller, more homogenous group than the boys. To obtain the data for this figure, I had my class get out their tape measures a second time and rule off the distance from the fingertips of the left hand to the fingertips of the right while the student they were measuring stood with arms outstretched like a big bird. After the assistant principal had come and gone (something about how the class was a little noisy, and although we were obviously having a good time, could we just be a little quieter), we recorded our results.

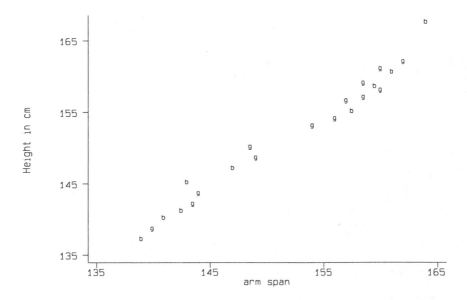

FIGURE 1.2. Arm Span in cm vs Height in cm for Students in Dr. Good's Sixth-Grade Class—by Sex.

Begin with a Picture

Charts and graphs have three purposes:

1. to summarize a set of observations
2. to compare two sets of observations (boys vs girls, U.S. vs CA.)
3. to stimulate our thinking.

These plots and charts have several purposes. One is to summarize the data. Another is to compare different samples or different populations (girls vs. boys, my class vs. your class). But their primary value is as an aid to critical thinking. The figures in this specific example may make you start wondering about the uneven way adolescents go about their growth. The exciting thing, whether you are a parent or as a middle-school teacher, is to observe how adolescents get more heterogeneous, more individual with each passing year.

1.2.2. From Observations to Questions

You may want to formulate your theories and suspicions in the form of questions: Are girls in the sixth grade taller on the average than sixth-grade boys (not just those in Dr. Good's sixth-grade class)? Are they more homogenous in terms of height? What is the average height of a sixth grader? How reliable is this estimate? Can height be used to predict arm span in sixth grade? At all ages?

You'll find straightforward techniques in subsequent chapters for answering these and other questions using resampling methods. First, I suspect, you'd like the answer to one really big question: Is statistics really much more difficult than the sixth-grade exercise we just completed? No, this is about as complicated as it gets.

Symbols are used throughout this book, so it helps if you've some experience with algebraic formulas. But you can make it successfully and enjoyably through the first seven chapters of this text even if you've never taken a college math course. If you're already knowledgeable in some academic discipline such as biology, physics, or sociology, you can apply what you're reading immediately. In fact, the best way to review and understand statistics is to take some data you've collected and/or borrowed from one of your textbooks and apply the techniques in this book. Do this now. Take a data set containing 20 or 30 observations. Find the minimum, median, and maximum. Make some charts, frequency distributions, and scatter plots. If you have kids or a younger sibling, pull a Tom Sawyer and see if you can get them to do it. Otherwise, get out your favorite spreadsheet or statistics package and start working through the tutorials.

1.2.3. Multiple Variables

The preceding results all dealt with cases in which we considered only one or two simultaneous observations. But in most practical studies we may need to consider and measure dozens, even hundreds. How then do we depict the results?

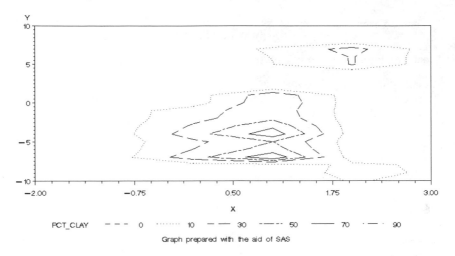

FIGURE 1.3. Clay Content as a Function of Location.

Capturing the Third Dimension

One way to see three variables on a single two-way plot is to use the third variable as a plotting symbol as in Figure 1.2. Another, used repeatedly in density estimation, a technique for classification described in Chapter 10, is to use contours that delineate equally spaced values of the third variable; an example is Figure 1.3, prepared with the aid of SAS™.* Still another is to use pseudo-perspective, as in Figure 10.8.

A fourth excellent solution is the scatterplot matrix. In Figure 1.4, we've used the Stata graph-matrix routine to plot all possible two-way scatterplots of four variables—median family income, housing units, median gross rent, and the population density. Each point corresponds to one of the fifty states.

1.2.4.   Contingency Tables

We can and should use tables to communicate our results. Hint: less is better.

Although St. John's is an Episcopalian school, my students came from a wide variety of religious backgrounds, as the following table reveals:

Episcopalian	Protestant	Catholic	Jewish	Other
6	5	6	2	3

We can use a similar sort of table to record the responses of my students to a poll made of their reaction to our new team approach.

Strongly Dislike	Dislike	O.K.	Liked	Like a Lot
2	3	5	5	7

---

*SAS is a product of SAS Institute Inc.

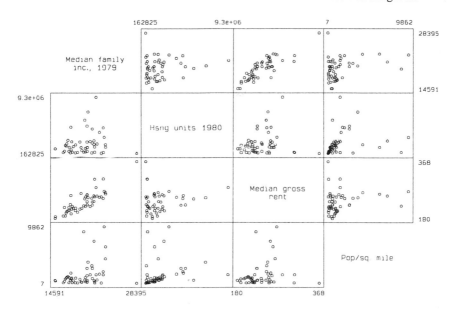

FIGURE 1.4. Matrix of Scatterplots.

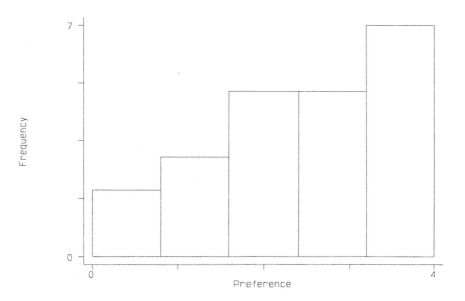

FIGURE 1.5. Histogram of Student Preferences.

I also recorded these results in a special form of bar chart known as a *histogram* in which the bars are contiguous. Note that in a histogram, the height of each vertical bar corresponds to the frequency or number in that ordered category.

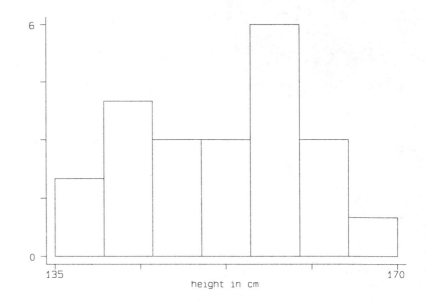

FIGURE 1.6. Histogram of Heights in Dr. Good's Class.

TABLE 1.3. Student height in cm.

137.5	142.5	147.5	152.5	157.5	162.5	167.5
2	4	3	3	6	3	1

We can also present the heights of my students in the form of a histogram (Figure 1.6) provided we first group the heights into intervals or categories as in *contingency table* 3.

How many categories? Freedman and Diaconis [1981] suggest we use an interval of length $2(IQR)n^{-1/3}$ where IQR stands for the interquartile range (see Section 1.2.1), and n is the sample size. For my class of $n = 22$ students with $IQR = 15$, the recommended length is approximately 10 cm, twice the length of the interval we used.[4]

A statistics package like Stata allows us to communicate this same information in the form of a one-way scatterplot (Figure 1.7), allowing us to quickly compare the boys and the girls.

1.2.5. Types of Data

Some variables we can observe and measure precisely to the nearest millimeter or even the nearest nanometer (billionth of a meter) if we use a laser measuring

---

[4]Regardless of what the formula says, we must always end up with an integral number of bins.

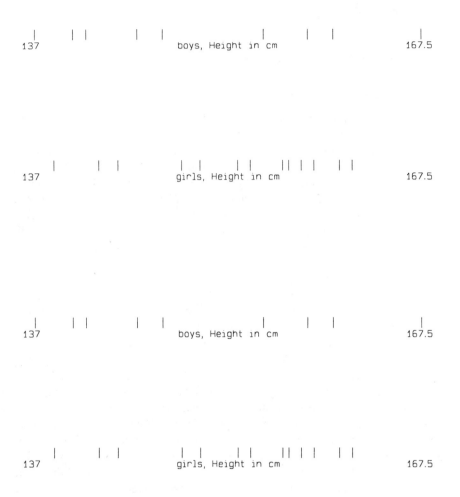

FIGURE 1.7. Scatterplots of Student Heights in Dr. Good's Class.

device. We call such data *continuous*. Some data, like that of the preceding section, fall into broad not-necessarily-comparable *categories*: males and females or white, black, and Hispanic. And some observations fall in between: they can be *ordered*, but they can't be added or subtracted the way continuous measurements can. An example would be the opinion survey in my classroom that yielded answers like "strongly favors," "favors," "undecided," "opposes," and "strongly opposes."

We can compute the mode for all three types of observation, and the median for both continuous and ordered variables, but the mean only makes sense for continuous data.[5]

1.3.   Measures of Location

1.3.1.   Median and Mode

Far too often, we find ourselves put on the spot, forced to come up with a one-word description of our results when several pages would do. "Take all the time you like," coming from my boss, usually means, "Tell me in ten words or less."

If you were asked to use a single number to describe data you've collected, what number would you use? Probably your answer is "the one in the middle," the *median* that we defined earlier in this chapter. When there's an odd number of observations, it's easy to compute the median. The median of 2, 2, 3, 4, and 5, for example, is 3. When there's an even number of observations, the trick is to split the difference: The median of the four observations 2, 3, 4, and 5 is 3.5, halfway between the two observations in the middle.

Another answer some people would feel comfortable with is the most frequent observation, or *mode*. In the sample 2, 2, 3, 4, and 5, the mode is 2. The mode is often the same as the median or close to it. Sometimes it's quite different and sometimes, particularly when there is a mixture of populations, there may be several modes.

Consider the data on heights collected in my sixth grade classroom, grouped as in Table 1.3. The mode is at 157.5 cm. But in this case, there may really be two modes, one corresponding to the boys, the other to the girls in the class. Often, we don't know in advance how many subpopulations there are, and the mode(s) serve a second purpose: to help establish the number of subpopulations.

Suppose you've done a survey of college students on how they feel about nudity on television with the possible answers being 0) never, 1) rarely, 2) if necessary, 3) O.K., and 4) more, more. Your survey results show 6 nevers, 4 rarelys, 5 if necessarys, 3 O.K.s, and 7 more, mores. The median is "necessary," the mode is "more, more." But perhaps there are really three modes, representing three populations, each centered at a different response (not being a sociologist, I won't attempt to describe the differences among the "nevers," the "if necessarys" and the "more mores"). We'll return to this topic in Chapter 10, where we'll use density estimation to help support our reasoning.

1.3.2.   Arithmetic Mean

The mean or arithmetic average is the most commonly used measure of location today although it can sometimes be misleading.

---

[5]A fourth data type, the binomial variable, which has exactly two categories, will be considered at length in the next chapter.

The Center of a Population

Median: the value in the middle; the halfway point; the value that has equal numbers of larger and smaller elements around it.

Arithmetic Mean or *arithmetic average*: the sum of all the elements divided by their number or, equivalently, the value such that the sum of the deviations of all the elements from it is zero.

Mode: the most frequent value; if a population consists of several subpopulations, there may be several modes.

The *arithmetic mean* is the sum of the observations divided by the number of observations. The mean of the sample 2, 2, 3, 4, 5, for example, is $(2 + 2 + 3 + 4 + 5)/5$ or 3.2. The mean height of my sample of 22 sixth-grade students is $(3335.5 \text{ cm})/22 = 151.6$ cm.

The arithmetic mean is a center of gravity, in the sense that the sum of the deviations of the individual observations about it is zero. This property lends itself to the calculus and is the basis of many theorems in mathematical statistics.

The weakness of the arithmetic mean is that it is too easily biased by extreme values. If we eliminate Pedro from our sample of sixth graders—he's exceptionally tall for his age at 5'7''—the mean would change to $3167/21 = 150.8$ cm. The median would change also, though to a much smaller degree, shifting from 153.5 to 153 cm.

1.3.3. Geometric Mean

The *geometric mean* is the appropriate measure of location when we are working with proportions or expressing changes in percentages, rather than absolute values. For example, if in successive months the cost of living was 110%, 105%, 110%, 115%, 118%, 120%, and 115% of the value in the base month, the average or geometric mean would be $(1.1 * 1.05 * 1.1 * 1.15 * 1.18 * 1.2 * 1.15)^{1/6}$.

1.4. Measures of Dispersion

As noted earlier, I don't like being put on the spot when I report my results. I'd rather provide a picture of my data—the cumulative frequency distribution, for example, or a one-way scatterplot—than a single summary statistic. Consider the following two samples:

Sample 1: 2, 3, 4, 5, 6 (in grams)
Sample 2: 1, 3, 4, 5, 7 (in grams)

Statistics and Their Algebraic Representation

Suppose we have taken n observations; let x_1 denote the first of these observations, x_2 the second, and so forth, down to x_n. Now, let us order these observations from the smallest $x_{(1)}$ to the largest $x_{(n)}$, so that $x_{(1)} \leq x_{(2)} \leq \ldots \leq x_{(n)}$.

A *statistic* is any single value that summarizes the characteristics of a sample. In other words, a statistic S is a function of the observations $\{x_1, x_2, \ldots, x_n\}$.

Examples include $x_{(1)}$, the minimum, and $x_{(n)}$, the maximum. If there's an odd number of observations, such as $2k + 1$, then $x_{(k+1)}$ is the median. If there's an even number of observations, such as $2k$, then the median is $(x_{(k)} + x_{(k+1)})/2$.

Let $x.$ denote the sum of the observations $x_1 + x_2 + \ldots + x_n$. We may also write $x. = \sum_{i=1}^{n} x_i$. Then the arithmetic mean $\bar{x} = x./n$.

The geometric mean is equal to $(x_1 * x_2 * \ldots * x_n)^{1/n}$.

The sample variance $s^2 = \sum_{i=1}^{n} (\bar{x} - x_i)^2/(n-1)$.

The standard deviation or L_2 norm s of the sample is the square root of the sample variance.

The first or L_1 sample norm $= \sum_{i=1}^{n} |\bar{x} - x_i|/(n-1)$.

Both have the same median, 4 grams, but it is obvious that the second sample is much more dispersed than the first. The span, or *range*, of the first sample is four grams, from a minimum of 2 grams to a maximum of 6. The values in the second sample range from 1 gram to 7; the range is six grams.

The classic measure of dispersion is the *standard deviation*, or L_2-norm, the square root of the sample *variance*, which is the sum of the squares of the deviations of the original observations about the sample mean, divided by the number of observations minus one. An alternative is the first or L_1-*norm*, the sum of the absolute values of the deviations of the original observations about the sample mean, divided by the number of observations minus one. For our first sample, the variance is $(4 + 1 + 0 + 1 + 4)/4 = 10/4$ grams^2, the *standard deviation* is 1.323 grams, and the L_1 norm is $(1 + 1 + 0 + 1 + 2)/4 = 1.25$ grams. For our second sample, the variance is $(9 + 1 + 0 + 1 + 9)/4 = 10$ grams^2, the standard deviation is 3.2 grams, and the L_1-norm is $(4 + 2 + 0 + 2 + 4)/4 = 3$ grams.

1.5. Sample versus Population

Generally, but not always, we focus our attention on a small sample of individuals drawn from a much larger population. Our objective is to make inferences about the much larger population from a detailed study of the sample.

We conduct a poll by phone of 100 individuals selected at random from a list provided by the registrar of voters, then draw conclusions as to how everyone in the county is likely to vote. We study traffic on a certain roadway five weekday mornings in a row, then draw conclusions as to what the traffic is likely to be during a similar weekday morning in the future.

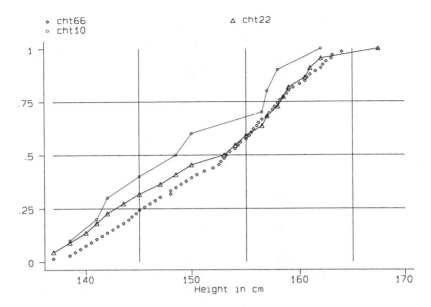

FIGURE 1.8. Cumulative Distribution Functions of Three Samples of Increasing Size Taken from the Same Population.

We study a sample rather than a population to both save time and cut costs. Sometimes, we may have no other alternative—when we are administering an experimental drug, for example, or estimating mean time to failure of a lightbulb or a disk drive. We don't expect the sample to exactly resemble the larger population—samples often differ widely from one another just as individuals do, but our underlying philosophy is that *the larger the sample, the more closely it will resemble the population from which it is drawn.*

Figure 1.8 illustrates this point with three closely related cumulative frequency distributions. The first is that of a random sample of ten students taken from my sixth-grade class at St. John's School, the second, depicted earlier, is that of the entire class, and the third is that of all the sixth-grade classes in the school. These three may be thought of as samples of increasing size taken from a still-larger population, such as all sixth-graders in Southern California.

Both theory and experience tell us that as the sample size increases, more of the samples will closely resemble the population from which they are drawn; the maxima and minima will get more extreme, while the sample median will get closer to the population median.

1.5.1. Statistics and Parameters

A *statistic* is any single value that summarizes some aspect of a sample. A *parameter* is any single value that summarizes some aspect of an entire population. Examples of statistics include measures of central tendency such as the sample mode, sample median, and sample mean; extrema such as the sample minimum and maximum; and measures of variation and dispersion such as the sample standard deviation and the sample L_1-norm. These same measures are considered parameters when they refer to the entire population rather than a sample.

1.5.2.   Estimating Population Parameters

The sample median provides us with an estimate of the population median, but it isn't the only possible estimate. We could also use the sample mean or sample mode to estimate the population median, or even the average of the minimum and the maximum. The sample median is the preferred estimate of the population median for two reasons: It is an unbiased estimate of the population median, that is, if we take a very large number of samples from the population of the same size as our original sample, the median of these sample medians will be the population median. And the sample median is consistent, that is, if we were to take a series of successively larger samples from the population, the sample median would come closer to the population median. In the next section, we shall consider a third property of this estimate: its consistency from sample to sample.

1.5.3.   Precision of an Estimate

We can seldom establish the *accuracy* of an estimate, for example, how closely the sample median comes to the unknown population median. But we may be able to establish its *precision*. That is, how closely the estimates derived from successive samples resemble one another.

   The straightforward way to establish the precision of the sample median, for example, is to take a series of samples from the population, determine the median for each of these samples and compute the standard deviation or the L_1-norm of these medians. Constraints on money and time make this approach unfeasible in most instances. Besides, if we had this many samples, we would combine them all into one large sample in order to obtain a much more accurate and precise estimate. One practical alternative, known as the *bootstrap*, is to treat the original sample of values as a stand-in for the population and to resample from it repeatedly, with replacement, recomputing the median each time.

   In Appendix 2, we've provided the details of the bootstrap algorithm for use with C, S+, Stata, and SAS. Our algorithm is general purpose and will prove effective whether we are estimating the variance of the sample median or some other, more complicated, statistic. If you don't have a computer, you could achieve the same result by writing each of the observations on a separate slip of paper, putting all the slips into an urn, and then drawing one slip at a time from the urn, replacing the slip in the urn as soon as you've recorded the number on it. You would repeat this procedure until you have a bootstrap sample of the same size as the original.

   As an example, our first bootstrap sample, which I've arranged in increasing order of magnitude for ease in reading, might look like this:

138.5	138.5	140.0	141.0	141.0	143.5	145.0	147.0	148.5	150.0
153.0	154.0	155.0	156.5	157.0	158.5	159.0	159.0	159.0	160.5
161.0	162.0

Several of the values have been repeated because we are sampling with replacement. The minimum of this sample is 138.5, higher than that of the original sample, the maximum at 162.0 is less, while the median remains unchanged at 153.5.

In the following second bootstrap sample, we again find repeated values; this time the minimum, maximum and median are 137.0, 167.5, and 148.5, respectively.

137.0	138.5	138.5	141.0	141.0	142.0	143.5	145.0	145.0	147.0
148.5	148.5	150.0	150.0	153.0	155.0	158.0	158.5	160.5	160.5
161.0	167.5

The medians of fifty bootstrapped samples drawn from our sample of sixth-graders ranged between 142.25 and 158.25, with a median of 152.75 (see Figure 1.9). They provide a feel for what might have been had we sampled repeatedly from the original population.

1.5.4.   Caveats

The sample is not the population; a very small or unrepresentative sample may give a completely misleading picture of the population as a whole. Take a look at the National League batting averages, particularly in the spring. You'll find quite a number of averages over .400 and many that are close to .100. Impossible, you say; there've been no .400+ hitters since Ted Williams, and anyone with a batting average close to .100 would be on his way to the minors. But it's not impossible if the player in question has only ten or twenty at bats. In the long run (that is, with a larger sample of at bats) that player will start to look more like we'd expect a major leaguer to look.

Administrators take note: If your institution only does a few of a particular procedure each year—a heart transplant, for example—don't be surprised if you

FIGURE 1.9. One Way Scatterplot of 50 Bootstrap Medians Derived from a Sample of Heights of 22 Students in Dr. Good's Sixth-Grade Class.

have results that are much higher or lower than the national average. It may just be a matter of luck rather than any comment on your facility.

The bootstrap will not reproduce all the relevant characteristics of a population, only those present in the sample, and cannot be used to estimate extrema such as maximum flood height or minimum effective dose.

The bootstrap can help us to estimate the errors that result from using a sample in place of the population (see Section 5.3), but one must be cautious in its application. Of the 50 bootstrap samples taken from my sample of 22 sixth-graders, 90% of the bootstrap medians lay between 147 and 156.5. But the probability that the population median lies between these values may be much larger or much smaller than 90%. Techniques for deriving more accurate interval estimates are described in Sections 5.6–5.8.

1.6.   Summary

In this chapter you learned to use summary statistics and graphs to report your findings. You learned definitions and formulas for measures of location and dispersion, and you learned to use sample statistics to estimate the parameters of the population from which the sample was drawn. Finally, you used the bootstrap to determine the precision of your estimates.

1.7.   To Learn More

Some of the best introductions to graphics for describing samples can be found in the manuals of such statistics packages as Stata and SAS. But see also the comprehensive reference works by Cleveland [1985, 1993] and Whittaker [1990]. Scott [1992] offers an excellent introduction to multivariate graphics. Sample data for self-study is included in Dalen [1992] and Werner et al. [1970].

The bootstrap has its origins in the seminal work of Jones [1956], McCarthy [1969], Hartigan [1969, 1971], Simon [1969] and Efron [1979, 1982]; Chapter 5 discusses its use in estimation in much more detail. One of its earliest applications to real-world data may be found in Makinodan et al. [1976]. Non-technical descriptions may be found in Diaconis and Efron [1983], Efron and Tibshirani [1991], Lunneborg [1985], and Rasmussen [1987].

1.8.   Exercises

1.8.1.   *Suggestions for Self-Study and Course Review*

Read over all of the exercises in the following section and be sure you understand what is required. Do several of the exercises using only a calculator. Use a spreadsheet or a statistics package such as Stata, S+, SAS, StatXact, or Testimate to complete whatever additional exercises you feel you need to do to feel comfortable with the methods you have learned. You may also download, without charge, the software I've developed, from my home page at **http://users.oco.net/drphilgood**.

Save your data on disk; we will return to these same data sets in future chapters. Make up and work through similar exercises using data that you or a colleague have collected.

1.8.2. Exercises

1. Are the following variables categorical, binomial, ordered, or continuous?
   Age   Age in years   Sex   Industry rank   Make of automobile
   Views on the death penalty   Annual sales   Shoe size   Method of
   payment (cash, check, credit card)
2. Use Table 1.4 to answer the questions and complete this exercise.

   a. How many observations are in this data set?
   b. How many variables does each observation comprise?
   c. Compute the mean, median, minimum, and maximum of each variable.
   d. What would you estimate the variance of your estimates of the mean and median to be?
   e. Plot a cumulative frequency distribution for each variable.
   f. Do you feel these distributions represent the aerospace industry as a whole? If not, why not?
   g. Make a scatter diagram(s) for the variables.

3. Nine young Republicans were asked for their views on the following two issues:

   a. Should a woman with an unwanted pregnancy have the baby even if it means dropping out of school and going on welfare?
   b. Should the government increase welfare support for people with children?

TABLE 1.4. Executive Compensation and Profitability for Seven Firms in the Aerospace Industry

Company	CEO Salary ($000s)	Sales ($000,000s)	Return on Equity (%)
Boeing	846	19,962.0	11.4
General Dynamics	1,041	11,551.0	19.7
Lockheed	1,146	12,590.0	17.9
Martin Marietta	839	6,728.5	26.6
McDonnell Douglas	681	15,069.0	11.0
Parker Hannifin	765	2,437.3	12.4
United Technology	1,148	19,057.1	13.7

For both issues, 10 = Strongly agree and 1 = Strongly disagree. Answers are recorded in the following table.

Woman's choice	9	8	1	2	2	10	10	4	3
Government	5	4	6	2	1	1	4	7	9

Draw a scattergram of these results. (We'll be referring to this example several times in succeeding chapters, so be sure to record your entries on disk.)

4. Make or buy a target divided into six or more sections and with three or more concentric circles and make 50 shots at the center of the target. (If you don't have a pistol or darts, mark off a piece of wrapping paper, put it on the floor, and drop coins on it from above.) Use a contingency table to record the number of shots, or coins, or darts in each section and each circle.

   a. How many shots (coins) would you expect to find in each section? Circle?
   b. How many shots (coins) did you find?
   c. Do you feel your shooting is precise? Unbiased? Accurate?

5. Use 50–100 bootstrap samples to estimate the standard deviation and the L_1-norm of the mean, median, and mode of the heights of my sample of 22 sixth-graders. Which is the least precise measure of central tendency? The most precise? How do these values compare with the standard deviation and the L_1-norm of the original sample?

6. Use Table 1.5 to answer the questions and complete this exercise.

   a. How many observations are there?
   b. How many variables are there per observation?
   c. Determine the minimum, median, and maximum for each variable, for each sex.
   d. Draw scatter diagrams that include information from all the variables.

7. During the 1970s, administrators at the U.S. Public Health Service complained bitterly that while the costs of health care increased over a four-year period from $43.00 to $46.50 to $49.80 to $53.70 per patient, their budget remained the same. What was the average cost per patient over the four-year period?

TABLE 1.5. Motrin Study 71: Site 8; Selected Variables from the Initial Physical

Subject ID	Sex	Age	Height	Weight
0118701	M	30	68	155
0118702	F	28	63	115
0118703	F	23	61	99
0118704	M	24	72	220
0118705	M	28	69	170
0118706	F	31	65	125
0118707	M	26	70	205

(Be careful: this question deals with costs per patient, not total costs or total patients. Which measure of central tendency should you use?)

8.

Country	Age		
	0	25	50
Argentina	65	46	24
Costa Rica	65	48	26
Dominican Republic	64	50	28
Ecuador	57	46	25
El Salvador	56	44	25
Grenada	61	45	22
Honduras	59	42	22
Mexico	59	44	24
Nicaragua	65	48	28
Panama	65	48	26
Trinidad	64	43	21

The preceding table records life expectancies at various ages for various Latin American countries.

a. Draw scatter plots for all pairs of variables.
b. Draw a contour plot.

9. Health care costs are still out of control in the 1990s, and there is wide variation in the costs from region to region. The Community Care Network, which provides billing services for several dozen PPOs, recorded the following cost data for a single dermatological procedure, debridement:

CIM	$198	200.2	242					
HAJ	83.2	193.6						
HAV	197							
HVN	93							
LAP	105							
MBM	158.4	180.4	160.6	171.6	170	176	187	
VNO	81.2	103.0	93.8					
VPR	154	228.8	180.4	220	246.4	289.7	198	224.4

a. Prepare both a histogram and a one-way scatterplot for this data.
b. Determine the median and the range.

10. In order to earn a C in Professor Good's statistics course, you should have an average score of 70 or higher. To date, your test scores have read 100, 69, 65, 60, and 60. Should you get a C?

11. My Acura Integra seems to be less fuel efficient each year (or maybe it's because I'm spending more time on city streets and less on freeways). Here are my miles-per-gallon result—when I remembered to record them—for the last three years:

1993	34.1	32.3	31.7	33.0	29.5	32.8	31.0	32.9		
1994	30.2	31.4	32.2	29.9	33.3	31.4	32.0	29.8	30.6	28.7
	30.4	30.0	29.1	31.2						
1995	30.1	30.1	29.5	28.5	29.9	30.6	29.3	32.4	30.5	30.0

a. Construct a histogram for the combined data.

b. Construct separate frequency distributions for each year.

c. Estimate the median miles per gallon for my automobile. Estimate the variance of your estimate.

12. The following table shows the results of a recent salary survey in the company I work for. As with most companies, only a few make the big bucks.

Wage ($000)	Number of employees
10-19	39
20-29	30
30-39	14
40-49	8
50-59	5
60-69	2
70-79	1
80-89	0
90-99	1

a. Create both a histogram and a cumulative frequency diagram.

b. What are the mean, median, and modal wages? Which gives you the best estimate of the average worker's take-home pay?

c. Which measure of central tendency would you use to estimate the costs of setting up a sister office in the northern part of the state?

13. The following vaginal titres were observed in mice 144 hours after inoculation with Herpes virus type II: 10,000; 9,000; 3,000; 2,600; 2,400; 1,700; 1,500; 1,100; 360, and 1.

a. What is the average value? (This is a trick question; the trick is to determine which of the averages we've discussed best represents this specific data set.)

b. Draw a one-way scatterplot and a histogram.
(Hint: This may be difficult unless you first transform the values by taking their logarithms; most statistics packages include this transform.)

14. Use the following table to answer the questions and complete this exercise.

LSAT	GPA	LSAT	GPA
545	2.76	594	2.96
555	3.00	605	3.13
558	2.81	635	3.30
572	2.88	651	3.36
575	2.74	653	3.12
576	3.39	661	3.43
578	3.03	666	3.44
580	3.07		

Is performance on the LSATs related to undergraduate GPA? Draw a scatter-gram for the data in the preceding table. Determine the sample mean, median, and standard deviation for each of the variables. Use the bootstrap to evaluate the precision of your estimate of the mean.

The preceding data is actually a sample of admission scores from a total of 82 American law schools. The mean LSAT for all 82 schools is 597.55. How does this compare with your estimate based on the sample?

15. **a.** If every student in my sixth-grade class grew five inches overnight, what would the mean, median, and variance of their new heights be?

b. If I'd measured their heights in inches rather than centimeters, what would the mean, median, and variance of their heights have been? (Assume that 1 cm = 2.54 inches.)

16. Summarize the billing data from four Swedish hospitals. (Hint: A picture can be worth several hundred numbers.) Be sure to create a histogram for each of the hospitals. To save you typing, you may download this data set from my home page at **http://users.oco.net/drphilgood**.

Hospital 1

64877	21152	11753	1834	3648	12712	11914	14290
17132	7030	23540	5413	4671	39212	7800	10715
11593	3585	12116	8287	14202	4196	22193	3554
3869	3463	2213	3533	3523	10938	17836	3627
30346	2673	3703	28943	8321	19686	18985	2243
4319	3776	3668	11542	14582	9230	7786	7900
7886	67042	7707	18329	7968	5806	5315	11088
6966	3842	13217	13153	8512	8328	207565	2095
18985	2143	7976	2138	15313	8262	9052	8723
4160	7728	3721	18541	7492	18703	6978	10613
15940	3964	10517	13749	24581	3465	11329	7827
3437	4587	14945	23701	61354	3909	14025	21370
4582	4173	4702	7578	5246	3437	10311	8103
11921	10858	14197	7054	4477	4406	19170	81327
4266	2873	7145	4018	13484	7044	2061	8005
7082	10117	2761	7786	62096	11879	3437	17186
18818	4068	10311	7284	10311	10311	24606	2427
3327	3756	3186	2440	7211	6874	26122	5243
4592	11251	4141	13630	4482	3645	5652	22058
15028	11932	3876	3533	31066	15607	8565	25562
2780	9840	14052	14780	7435	11475	6874	17438
1596	10311	3191	37809	13749	6874	6874	2767
138133							

Hospital 2

4724	3196	3151	5912	7895	19757	21731	13923
11859	8754	4139	5801	11004	3889	3461	3604
1855							

Hospital 3

4181	2880	5670	11620	8660	6010	11620	8600
12860	21420	5510	12270	6500	16500	4930	10650
16310	15730	4610	86260	65220	3820	34040	91270
51450	16010	6010	15640	49170	62200	62640	5880
2700	4900	55820	9960	28130	34350	4120	61340
24220	31530	3890	49410	2820	58850	4100	3020
5280	3160	64710	25070				

Hospital 4

10630	81610	7760	20770	10460	13580	26530	6770
10790	8660	21740	14520	16120	16550	13800	18420
3780	9570	6420	80410	25330	41790	2970	15720
10460	10170	5330	10400	34590	3380	3770	28070
11010	19550	34830	4400	14070	10220	15320	8510
10850	47160	54930	9800	7010	8320	13660	5850
18660	13030	33190	52700	24600	5180	5320	6710
12180	4400	8650	15930	6880	5430	6020	4320
4080	18240	3920	15920	5940	5310	17260	36370
5510	12910	6520	5440	8600	10960	5190	8560
4050	2930	3810	13910	8080	5480	6760	2800
13980	3720	17360	3770	8250	9130	2730	18880
20810	24950	15710	5280	3070	5850	2580	5010
5460	10530	3040	5320	2150	12750	7520	8220
6900	5400	3550	2640	4110	7890	2510	3550
2690	3370	5830	21690	3170	15360	21710	8080
5240	2620	5140	6670	13730	13060	7750	2620
5750	3190	2600	12520	5240	10260	5330	10660
5490	4140	8070	2690	5280	18250	4220	8860
8200	2630	6560	9060	5270	5850	39360	5130
6870	18870	8260	11870	9530	9250	361670	2660
3880	5890	5560	7650	7490	5310	7130	5920
2620	6230	12640	6500	3060	2980	5150	5270
16600	5880	3000	6140	6790	6830	5280	29830
5320	7420	2940	7730	11630	9480	16240	2770
6010	4410	3830	3280	2620	12240	4120	5030
8010	5280	4250	2770	5500	7910	2830	11940
9060	20130	10150	6850	10160	7970	12960	31550

17. Discuss the various reasons one might take a sample rather than examine an entire population. How might we tell when a sample is large enough?

CHAPTER 2

Cause and Effect

In this chapter, you'll develop formal models linking cause and effect. You'll begin with your reports, listing and, preferably, graphing your anticipated results. You'll use these graphs to derive formal models combining deterministic and stochastic (random) elements. Probability and distribution theory help you draw the representative samples you need to assess your models.

2.1. Picturing Relationships

Begin with your reports. List your expectations and, preferably, express them in the form of charts and graphs.

Picturing a relationship in physics is easy. A funeral procession travels along the freeway at a steady 55 miles an hour, so that when we plot its progress on a graph of distance traveled versus time, the points all fall along a straight line as in Figure 2.1.

A graph of my own progress when I commute to work looks a lot more like Figure 2.2; this is because I occasionally go a little heavy on the gas pedal, while other times traffic slows me down. Put a highway patrol car in the lane beside me and a graph of my progress would look much like that of an army convoy or a funeral procession. The underlying pattern is the same in each case—the distance traversed increases with time. But in real life, random fluctuations in traffic result in accelerations and de-accelerations in the curve.

The situation is similar but much more complicated when we look at human growth (Figure 2.3). A rapid rate of growth for the first year, slow steady progress for the next 10–14 years and then, almost overnight it seems, we're a different size. Inflection points in the growth curve are different for different individuals. My wife was 5'6" tall at the end of grade six, and 5'7" tall when she married me.

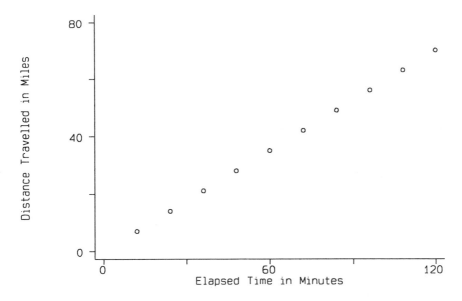

FIGURE 2.1. Graphing the Progress of a Funeral Procession.

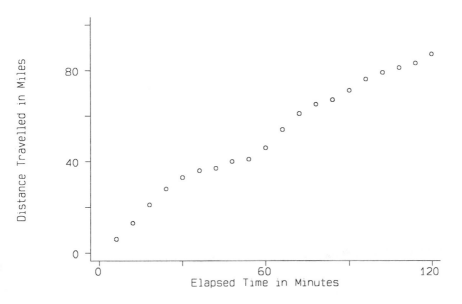

FIGURE 2.2. Graphing my Progress Through Freeway Traffic.

I was 4'10" at the end of grade six, 5' at the beginning of grade ten, and 5'10" when she married me. The pattern was the same for both of us, but the timing of our growth spurts differed.

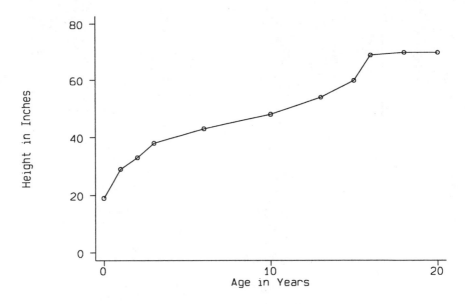

FIGURE 2.3. Growth Curve of Donny Travaglia.

These differences among individuals are why we need a way to characterize both average behavior and variation. A single number just won't do.

Before I launch an investigation, I start thinking about how and in what form I will report my results. I try to imagine the relationships I will be depicting. Suppose, for example, I'm planning to investigate the relationship between education and income. What variables should I use to quantify this relationship? Years of education? Annual income at age 30? Total lifetime income? To gain further insight, I'll pretend I've done a preliminary survey of several hundred individuals. I write down guesstimates of average responses and transfer these guesstimates to a two-way plot as in Figure 2.4.a.

A positive but nonlinear relationship is depicted in Figure 2.4.a with income rising as one completes high school and college and falling off again with each year beyond the first in graduate school. Other possibilities, depicted in Figure 2.4. b, c, and d, include a positive linear relationship, a negative linear relationship, and no relationship between income and education.

My next step is to begin to quantify this relationship in the form of an equation. Figures 2.1, 2.2, 2.4. b and c, and the initial rising portion of Figure 2.4. a all have the same underlying linear form: $Y = a + bX$, where Y is the dependent variable (income or distance traveled) X is the independent variable (years of education or time) and a and b are the to-be-estimated intercept and slope of the line.

Suppose for example, we were to write $I = \$20,000 + \$5,000E$ where I stands for income and E for years of education after high school. This relationship is depicted in Figure 2.4.b. Among its implications are that without college ($E = 0$)

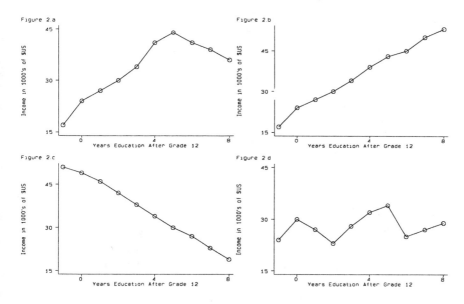

FIGURE 2.4. Various Trend Lines for Income as a Function of Years of Education: a) Expected, b) Rising Linear, c) Falling Linear, d) No Association.

Developing and Testing a Model

1. Begin with the reports. Picture relationships.
2. Convert your pictures to formal models with both deterministic and stochastic elements.
3. Plan your sampling method. Define the population of interest. Ensure that a) the sample is representative and b) the observations are independent.

---

average annual income is \$20,000, while with college completed ($E = 4$) average income is doubled, or \$40,000.

2.2. Unpredictable Variation

If our model were perfect, then all the points would lie exactly on a straight line. We might be able to improve the fit by writing the dependent variable, income in the previous example, as a function of several different variables X_1, X_2, X_3 each representing some characteristic that might influence future income, but, again, it's unlikely the fit would be perfect. Even with time-tested models of the kind studied in freshman science laboratories, the observations just won't cooperate. There always seems to be a portion we can't explain, that is, the result of observer error,

TABLE 2.1. Registrants and Servings at Fawlty Towers

Registrants	Maximum Servings	Registrants	Maximum Servings
289	235	339	315
391	355	479	399
482	475	500	441
358	275	160	158
365	345	319	305
561	522	331	225

or DNA contamination, or a hundred other factors we did not think of measuring or were unable to measure.

For simplicity, let's represent all the different explanatory variables by the single letter X, and again suppose that even with all these additional variables included in our model, we still aren't able to predict Y exactly. There is still a small fraction ϵ of each observation that we can't explain. Our solution is to write Y as a mixture of deterministic and stochastic (random) components,

$$Y = bX + \epsilon,$$

where X represents the variables we know about, b is a vector of constants used to apportion the effects of X, and ϵ denotes the error or random fluctuation, the part of the relationship we can't quite pin down or attribute to any specific cause.

2.2.1.   Building a Model

Imagine you are the proud owner of Fawlty Towers and have just succeeded in booking the International Order of Arcadians and Porcupine Fanciers for a weekend conference. In theory, you should prepare to serve as many meals as the number of registrants, but checking with your fellow hotel owners, you soon discover that with no-shows and nondiners you can get by with a great many less. Searching through the records of the former hotel owners, you come up with the data in Table 2.1. You convert these numbers to a graph (Figure 2.5) and discover what looks almost like a linear relationship!

Quickly, you leap ahead to Chapter 5.3 and use a method found there for estimating the parameters of the line:

$$\text{Servings} = .94 * \text{Registrants} - 20.55001$$

Of course, there are discrepancies (also known as *residuals*), as can be seen from Figure 2.6 and as you note in the following table:

Is there an explanation for these discrepancies? There were 331 registrants, 290 predicted meals, and only 225 meals served! Well, one thing you forgot to mention to the officers of the IOAPF before you signed them up is that for most

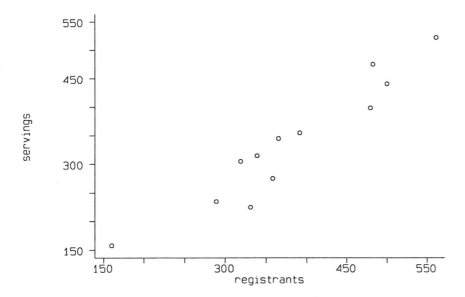

FIGURE 2.5. Scatterplot of Servings vs Conference Registrants at Fawlty Towers

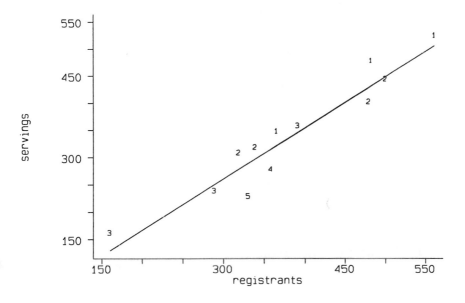

FIGURE 2.6. Regression of Servings on Conference Registrant at Fawlty Towers.

of the year the weather out Fawlty Towers way is awful.[1] Perhaps bad weather is

---

[1] The pictures you showed them of Fawlty Towers, taken on your own Hawaiian vacation, didn't help the situation. What you didn't realize and will be forced to deal with later when

the explanation. Digging deeper into the previous owners' records, you come up with the expanded Table 2.2.

| | Servings | | |
Registrants	Actual	Predicted	Residual
289	235	251	−15.9
391	355	347	8.3
482	475	432	42.8
358	275	316	−40.7
365	345	322	22.7
561	522	506	15.6
339	311	298	17.1
479	399	429	−30.4
500	441	449	−8.1
160	158	130	28.3
319	305	279	25.9
331	225	290	−65.4

Adding a term to your model to account for the weather and again applying the methods of Chapter 4 yields the following formula:

$$\text{Servings} = .78 * \text{Registrants} - 27 * \text{Weather} + 104$$

The residuals, as you can see from Table 2.3, are much smaller, but the fit still is not perfect.

2.3.   Two Types of Populations

In Chapter 1, we studied samples taken from real populations, that is, populations of actual individuals we could reach out and touch. But in many scientific studies, the populations we examine are only hypothetical. Examples include potential registrants for the IOAPF conference, women 25–40 years of age who might someday take aspirin, the possible outcomes of 25 successive coins tosses. Our choice of

TABLE 2.2. Servings as a Function of Registants and Weather

Registrants	Servings	Weather	Registrants	Servings	Weather
289	235	3	339	315	2
391	355	3	479	399	2
482	475	1	500	441	2
358	275	4	160	158	3
365	345	1	319	305	2
561	522	1	331	225	5

you come up 100 meals short is that the weather in Nova Scotia, where most of those Arcadians live, is awful for most of the year also.

TABLE 2.3. Residuals from First Model

Registrants	Actual	Servings Predicted	Residual
289	235	250	−14.5
391	355	329	25.6
482	475	455	20.2
358	275	277	−1.6
365	345	363	−18.1
561	522	517	5.3
339	315	316	−0.7
479	399	425	−26.4
500	441	442	−0.9
160	158	148	9.6
319	305	300	4.9
331	225	228	−3.4

words—potential, possible, might be—confirms the distinction. The good news is that we can characterize these hypothetical populations in much the same way we characterize the real ones, by specifing means, medians, standard deviations, and percentiles.

The reason we need to consider hypothetical populations is that all our observations have a stochastic component. In some instances, for example, when we repeatedly measure a student's height with a ruler, the random component is inherent in the measuring process, so-called *observation error*; in others, the variation is part of the process itself. What will be the effect of my taking an aspirin? The answer depends on my current state of health, my current body temperature, even the time of day. Broaden the scope of the question; ask, for example, what the effect of a single aspirin will be on an individual selected at random, and we introduce a dozen more sources of variation, including age and sex.

2.3.1. Predicting the Unpredictable

Many of these and similar observations we make in practice seem to have so many interacting causes that they are entirely unpredictable. Examples include the outcome of a flip of a coin or the roll of a die, the response to a medication, or the change in the price of a commodity between yesterday and today. The possible outcomes may be equally likely, as in the flip of a coin, or weighted in favor of a specific outcome as in "Take two aspirins and you'll probably feel better in the morning." Even in this latter case, some patients won't get better, and others will acquire additional symptoms resulting directly from the medication.

Although we may not be able to predict the outcome of an individual trial, we may be able to say something about the outcome of a series of trials. In the next section, we'll consider the simplest such trials, ones with exactly two possible outcomes.

In the Long Run: Some Misconceptions

When events occur as a result of chance alone, anything can—and usually will— happen. You roll craps seven times in a row, or you flip a coin ten times and ten times it comes up heads. Both these events are unlikely, but they are not impossible. Before reading the rest of this section, test yourself by seeing if you can answer the following:

You've been studying a certain roulette wheel that is divided into 38 sections, 1–36, 0, and 00 for more than four hours and not once during those four hours of continuous play has the ball fallen into the number 6 slot. You conclude that (1) number 6 is bound to come up soon and bet on it; (2) the wheel is fixed so that number 6 will never come up; or 3) the odds are exactly what they've always been and in the next four hours number 6 will probably come up about 1/38th of the time.

If you answered (2) or (3), you're on the right track. If you answered (1), think about the following equivalent question:

You've been studying a series of patients treated with a new experimental drug, all of whom died in excruciating agony despite the treatment. You (1) conclude the drug is bound to cure somebody sooner or later and take it yourself when you come down with the symptoms, or (2) decide to abandon this drug and look for an alternative.

2.4. Binomial Outcomes

The simplest stochastic variable is the binomial trial—success or failure, agree or disagree, heads or tails. If the coin is fair, that is, if the only difference between the two outcomes lies in their names, then the probability of throwing a head is 1/2, and the probability of throwing a tail is also 1/2. (By definition, the probability that something will happen is 1, the probability that nothing will occur is 0; all other probabilities are somewhere in between.[2])

What about the probability of throwing heads twice in a row? Ten times in a row? If the coin is fair and the throws are independent of one another, the answers are easy: $1/4 = 1/2 x 1/2$ and $1/1,000 = 1/2 \times 1/2 x \cdots 1/2 = (1/2)^{10}$.

These answers are based on our belief that when the only difference among several possible outcomes are their labels, "heads" and "tails," for example, the various outcomes will be equally likely. If we flip two fair coins or one fair coin twice in a row, there are four possible outcomes HH, HT, TH, and TT. Each outcome has equal probability of occurring. The probability of observing the one outcome in which we are interested is 1 in 4 or 1/4th. Flip the coin ten times and

---

[2] If you want to be precise, the probability of throwing a head is probably only 0.49999, and the probability of a tail is also only 0.49999. The leftover probability of 0.00002 is the probability of all the other outcomes—the coin stands on edge, a sea gull drops down out of the sky and takes off with it, and so on.

there are 2^{10}, or 1,000 possible outcomes; one such outcome might be described as HTTTTTTTTH.

Unscrupulous gamblers have weighted coins so that heads comes up more often than tails. In such a case, there is a real difference between the two sides of the coin and the probabilities will be different from those just described. Suppose that as a result of weighting the coin the probability of getting a head is now p, where $0 \leq p \leq 1$, and the complimentary probability of getting a tail, (or not getting a head) is $1 - p$. (Again, $p + (1 - p) = 1$.) Again, we ask, what is the probability of getting two heads in a row? The answer is p^2.

Here's why: To get two heads in a row, we must first throw a head. In others words, only a fraction p of the trials are of interest to us. The proportion $1 - p$ in which we throw a tail initially is no longer of interest. Of the fraction p of two successive trials that begin with a head, only a further fraction p will also end with a head, that is, only $p \times p = p^2$ trials result in HH. Similarly, the probability of throwing ten heads in a row is p^{10}.

By the same line of reasoning, we can show the probability of throwing nine heads in a row followed by a tail using the same weighted coin each time is $p^9(1 - p)$. What is the probability of throwing nine heads in ten trials? Is it also $p^9(1 - p)$? No, this latter event includes the case in which the first trial is a tail and all the rest are heads, the second trial is a tail and all the rest are heads, the third trial is ... , and so on. There are ten different ways in all. These different ways are mutually exclusive. The probability of the overall event is the sum of the individual probabilities, or $10p^9(1 - p)$.

What is the probability of throwing exactly five heads in ten tosses of a coin? The answer to this question requires that we understand something about permutations and combinations, a concept that will be extremely important in later chapters.

2.4.1. Permutations and Combinations

Suppose we have three horses in a race. Call them A, B, and C. A could come in first, B could come in second, and C would be last. ABC is one possible outcome, or *permutation*. But so are ACB, BAC, BCA, CAB, and CBA. There are six possibilities in all. Now suppose we have a nine-horse race. We could write down all the possibilities, or we could use the following trick: We choose a winner (nine possibilities); we choose a second place finisher (eight remaining possibilities), and so on until all positions are assigned—a total of $9! = 9 \times 8 \times 7 \times 6 \times 5 \times 4 \times 3 \times 2 \times 1$ possibilities in all.

Normally, in a horse race, all our attention is focused on the first three finishers. How many possibilities are there? Using the same reasoning, it is easy to see that there are $9 \times 8 \times 7$ possibilities or $9!/6!$.

Suppose we ask a slightly different question: In how many different ways can we select three horses from nine entries without regard to order (that is, we don't care which comes first, which second, or which third). In the previous example, we distinguished between first, second, and third place finishers; now we're saying the order of finish doesn't make a difference. We already know there are $3! =$

Permutations vs. Rearrangements

There are $10! = 10 \times 9 \times 8 \times 7 \times 6 \times 5 \times 4 \times 3 \times 2$ different ways we can arrange ten items in a line. But there are only $\binom{10}{5} = \frac{10!}{5!5!}$ different ways we can divide ten items among two groups of size five.

$3 \times 2 \times 1 = 6$ different permutations of the three horses that finish in the first three places. So we take our answer to the preceding question $9!/6!$ and divide this answer in turn by $3!$. We write the result as $\binom{9}{3}$, which is usually read as *9 choose 3*.

Note $\binom{9}{6} = \binom{9}{3}$.

2.4.2. Back to the Binomial

We used horses in the preceding example, but the same reasoning can be applied to coins or survivors in a clinical trial.[3] What is the probability of five heads in ten tosses? What is the probability that five of ten breast cancer patients will still be alive after six months?

We answer this question in two stages. First, what is the number of different ways we can get five heads in ten tosses? We could have thrown HHHHHTTTTT or HHHHTHTTTT, or some other combination of five heads and five tails for a total of 10 choose 5, $\binom{10}{5}$ or $10!/(5! \times 5!)$ ways. The probability that the first of these events occurs—five heads followed by five tails—is $(1/2)^{10}$. Combining these results yields

$$\Pr\{5 \text{ heads in 10 throws of a fair coin}\} = \binom{10}{5}(1/2)^{10}.$$

We can generalize the preceding to an arbitrary probability of success p, $0 \leq p \leq 1$. The probability of failure is $1 - p$. The probability of k successes in n trials is given by the binomial formula

$$\binom{n}{k} p^k (1 - p)^{n-k} \text{ for } 0 \leq k \leq n.$$

Figure 2.7 depicts the probability of k or fewer successes in 10 trials for various values of p. The expected number of successes in many repetitions of n binomial trials, each of which has a probability p of success, is np; the expected variance is $np(1 - p)$.

---

[3] If, that is, the probability of survival is the same for every patient. When there are obvious differences from trial to trial—one is an otherwise healthy 35-year-old male, the other an elderly 89-year-old who has just recovered from pneumonia, this simple binomial model would not apply; see also Section 2.6.1

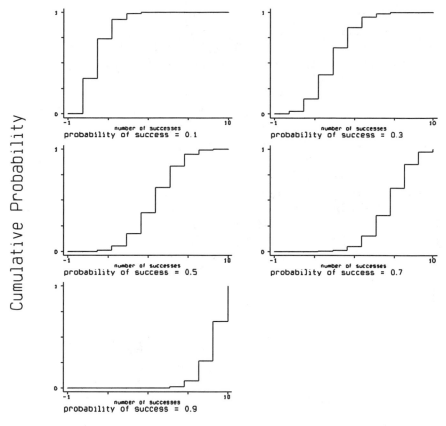

FIGURE 2.7. Cumulative Probability Distribution for the Binomial with 10 Trials for Various Values of p from 0.1 to 0.9.

2.4.3.   Probability

We have already discussed probability in an informal way: the purpose of this section is to summarize our previous discussion. Let $\Pr(A)$ denote the probability that an event A occurs, for example, a toss of a coin yields heads, an advertisement for your product is read by at least 1,000 people, or a patient who takes two aspirin feels better within the hour.

$$0 \leq \Pr(A) \leq 1.$$

An impossible event has probability 0; a sure thing has probability 1. Most of the other rules of probability are equally obvious.

Either something happens, or it doesn't: $\Pr(A) + \Pr(\text{not } A) = 1$. The probability that a fair coin comes up heads is $1/2$; the probability it comes up tails is $1 - 1/2 = 1/2$. More generally, if we have k equally likely mutually exclusive events, that is, the only distinction among them is their labels, then each will occur with probability $1/k$. Consider, for example, the probability of throwing a 6 with a six-sided die or that "Whistler's Brother" wins, when he is one of ten horses in a race in which each animal has been perfectly handicapped by the addition of weights. (In the latter case, a bet on "Whistler's Brother," at less than 10–1 odds is an indication that you feel you are a better handicapper than the person who specified the weights.)

If B is a collection of events that includes A $Pr(B) \geq Pr(A)$.

If C is a collection of events $\{B_1, B_2, \cdots, B_n\}$ that are mutually exclusive, then $Pr(C) = Pr(B_1) + Pr(B_2) + \cdots Pr(B_n)$. An example of such a collection is the event C: "throws heads at least four out of five times in a row," which is made up of two mutually exclusive events, B_1, "throws heads four out of five times in a row," and B_2, "throws heads five out of five times in a row." These events are mutually exclusive because only one of them can occur. The latter consists of the single event HHHHH, but the former is actually composed of five separate events HHHHT, HHHTH, HHTHH, HTHHH, and THHHH. As there are $2^5 = 32$ distinct possibilities in all, $Pr(B_2) = 1/32$, $Pr(B_1) = 5/32$, and $Pr(C) = 6/32$.[4]

2.5. Independence

Underlying most statistical procedures is the assumption that successive observations are independent of one another. The sequence of binomial trials described earlier is just one example of many. Violate this assumption, and resampling methods such as the bootstrap and permutation tests are no longer applicable.

Independence means all of the following:

1. Independent trials do not influence each other's outcome.
2. If we know the outcome of one of the trials, it won't provide us with additional insight into the possible outcome of the other trials.
3. The probability of the joint outcome of two independent trials can be written as the product of the probabilities of the individual trials. (We took advantage of this multiplication rule in Section 2.4.)

An example of independent events is: A—the coin came up heads and B—the person residing at 1063 Elm Street votes Republican.

The best way to understand the concept of independence is to consider a few other simple examples. You flip ten coins once each. Are the ten outcomes independent? Of course. You ask the first ten people you meet in the street if they are against compulsory pregnancy; are their answers independent? Maybe. Or maybe the ten just walked out of a meeting of young ditto heads.

---

[4] Actually, each of these so called "fundamental events" is made up of still more detailed possibilities, such as "the first time I threw the coin in the air, it spun around five times, was touched briefly by a wind current, bounced an inch above my hand, bounced a second time a mere fraction of an inch, and then came to rest head up. . . ," but you get the idea.

What about shuffling cards? Shuffle the cards thoroughly and successive deals are independent. Shuffle them carelessly and the beginning of one card game will bear a strong resemblance to the finish of the last.

Measure the blood pressures of two different individuals and the values are independent, right? Maybe. If the measurements were taken by an experienced individual in a quiet room, they would be. But walk into a physiology class some day when the class is just learning to take blood pressure. Everybody is hyper, and all the measurements tend to be on the high side.[5]

What about two successive measurements taken on the same individual such as blood pressures or test scores? Obviously these measurements are dependent. But what about the first observation and the difference between it and the next measurement? These two may be independent. If the Dow-Jones index is high today, we're probably safe in believing it will be on the high side tomorrow, but we don't know whether it will go up or down in the interval or by how many points.

2.6.   Selecting the Sample

To evaluate your tentative theory, you plan to draw a sample from a much larger population and then apply resampling methods to estimate the characteristics of the population from which the sample is derived. But will the sample you draw be representative?

As we saw in Section 2.3, the word population has a much broader meaning in statistics than in ordinary speech. Sometimes the word refers to a group of distinguishable individuals, such as all sixth-graders in Southern California or all registered voters in the British Isles. In other instances it may refer to a set of repetitions—every time I flip this coin, or every time a leukemia patient is treated with cyclohexyl–chloroethyl nitrosurea.

It's easy to get the two meanings confused. A psychologist friend of mine, a very bright fellow, succeeded in training a mouse to run a maze, and then in running that same mouse—her name was Mehitabel—through some 100 planned variations of the basic course. "A sample of size 100, that's a lot. A lot of work, too," he confided.

Was it a large sample? He had run 100 variations of an experiment with a mouse named Mehitabel. And he had obviously developed an extensive theory about Mehitabel's maze-running behavior. But as far as the general population of mice went, "You have a sample of size one," I told him.

Fortunately, he was able to use the concepts advanced in Chapter 7 of this book to design a more efficient experiment in which ten different mice were each run through the same ten mazes.

---

[5] Such measurements are said to be conditionally independent, given the room.

Representative Samples

1. Observations are independent.
2. Each item in the population has equal probability of being selected.
3. Identical method of measurement is used on every item.
4. Observations are made in random order.

He did have to consult me again before he came up with the final design. The first time around, his idea was to repeat the initial experiment using Mehitabel's nine littermates. Do you see what's restrictive about this idea?

The same restriction would arise if one did a survey of political preferences by going from house to house interviewing the occupants while they sat around the dinner table. Their answers would not be the same as those expressed in the privacy of the polling booth. People care what others think, and when asked for an opinion about an emotionally charged topic in public, may or may not tell the truth. In fact, they're very unlikely to tell the truth if they think third parties are listening in. If a wife hears a husband say "Democrat," she may just reply "Democrat" herself to avoid an argument. To obtain a true representative sample, our observations must be independent of one another.

2.6.1. A Young Wo_Man Tasting Herbal Tea

A young woman burst into my office last Tuesday evening, her nostril flaring.[6] "I can identify the correct brand of herbal tea five times out of five," she said.

"You're on," I said. "And I'll buy if you succeed."[7]

We went to the Student Union, where I proceeded to turn the whole coffee shop upside down searching for five matching cups. "What's the difference?" I was asked. The difference is that I wanted this to be an experiment about herbal tea, not about cups. I also made sure I had enough hot water for all five cups, and that all five brands were caffeine free, not mixtures of tea and herbs. Finally, I asked a second student if he wouldn't mind administering the test while I waited in the next room.

"Whatever for?" he wanted to know. The answer to this question is that I didn't want my wizard of the tea bags to be able to discern from the expression on my face whether she was on the right track. I would prepare the tea in one room, remove the tea bags, and hand the cups to the second student. He would carry the cups to her and record her answers. Oh, and I had him stick a blindfold on her while I was in the next room preparing the tea.

---

[6]The other nostril was pierced by a large metal safety pin.

[7]Friends (?) of mine say it's extremely unlikely I ever made even this qualified offer to buy a round.

Do you feel this is a lot of effort to ensure the experiment measured exactly what she said it would? Not at all. I would have had to pay for all five cups if she'd gotten it right.[8]

2.6.2.   Random Sampling and Representative Samples

An individual sample may not always be *representative*, that is, an exact copy of the population, but the method by which you sample must be representative in the sense that it ensures that every possible sample is equally likely.

Random sampling would seem essential in the selection of jurors. The California Trial Jury Selection and Management Act[9] states "It is the policy of the State of California that all persons selected for jury service shall be selected at random from the population of the area served by the court; that all qualified persons have an equal opportunity . . . to be considered for jury service in the state and an obligation to serve as jurors when summoned for that purpose; and that it is the responsibility of jury commissioners to manage all jury systems in an efficient, equitable, and cost-effective manner in accordance with this chapter."

Are extremes, such as a jury of 12 white males, or the 10 blacks, 1 white, and 1 Hispanic who served on the O.J. Simpson jury, a legal inequity?

Random is not the same as fair or equitable. Anything can happen—and often does—when samples are selected at random. Red appears ten times in ten consecutive spins of the roulette wheel. Impossible? Not at all. A result like this (or one equally improbable, like getting black ten times in a row) would be expected once in every 500 times (see Section 2.2).

Justice Blackmun notes in *Ballew v. Georgia* (1978)[10] that if a minority group comprises ten percent or less of a population, a jury of 12 persons selected at random from that population will fail to contain members of that minority at least 28% of the time.[11]

In *People v. Charles Manson* (1977)[12], the court held that defendants were not entitled to a jury of any particular composition, nor was there a requirement that the petit jury be representative of various distinct economic, political, social, or racial groups in the community.

The court reached a similar conclusion in *O'Hare v. Superior Court* (1987)[13], saying "The 6th Amendment does not entitle a defendant to a venire drawn from or representative of the entire county. What the 6th Amendment does guarantee is a jury drawn from a venire from which no member of the local community was arbitrarily or unnecessarily excluded."

---

[8]She got one out of five right, about what you'd expect by chance alone.

[9]Title 3, C.C.P. Section 191.

[10]435 U.S. 223, 236–37, 98 S.Ct.1029, 55 L.Ed.2d 234.

[11]Turning to Section 2.2, you can see that Justice Blackmun definitely deserves an A in statistics.

[12]71 Cal. App. 3d 1, 139 Cal. Rptr. 275, certiorari denied *Manson v. California* (1978) 435 U.S. 953, 98 S.Ct. 1582, 55 L.Ed. 2d 803.

[13]43 Cal. 3d 86, 729 P. 2d 766, 233 Cal. Rptr 332

We must apply the same principles when we select our samples as jury commissioners are bound to when they select their panels.[14]

Can't we bend the rules a little? Do away with that blindfold or let the person recording the observations know which cup of tea is which? The courts won't allow it, and neither should we.

In *U.S. v. Northside Realty Assoc.* (5th Cir 1981)[15], the court dismissed indictments handed down by the grand jury, finding there had been a substantial failure in the mechanism by which the grand jury was selected. This mechanism, common today now that most courts are computerized, consists of three steps: First, a master jury panel is selected in accordance with statute; second, the number of the first juror, the random-number seed, is selected; and third, a quasi-random computer number generator is used to select all other jurors.

The first and third steps were not at issue in *Northside Realty*. At step two, the court clerk had selected a "convenient" number. Inspection of past grand juries revealed that the clerk had used the same starting seed repeatedly, and hence, such being the nature of computerized quasi-random number generators, the numbers of the other "randomly selected" jurors had also been repeated.

Not all deviations from random selection result in reversal, nor should they. In *People v. Viscotti* (1992)[16], the issue was whether the trial court erred in taking the first 12 jurors from the panel, rather than selecting 12 at random. The court held that a material departure from statutory procedures had not occurred as the panel itself had been selected at random from the population.

Such a decision may not be correct in other contexts. For example, repeated studies have shown that the first rodents pulled from a cage tend to be more active and have higher cortisone levels than the others. In some experiments, this would introduce a bias.

Our objective in selecting a sample should be to control, measure, or eliminate as many extraneous factors as we can and then to randomize our selection with whatever is left (see Chapter 8).

2.7.   Summary

In this chapter, you learned to express cause-effect relationships in the form of both a graph and a function. You learned that most cause-effect relationships have both deterministic and stochastic components. You were introduced to the concept of the binomial trial and to probabilities, permutations, and combinations. You studied the concept of independence and learned the importance of drawing samples that would be representative of the underlying population.

---

[14] An acceptable alternative, discussed in Section 7.1.1, is when we take separate random samples from each of several different subgroups in a population—ten nonsmoking white males and five black females who are heavy smokers, for example—and combine them all into a single large sample.

[15] 659 F. 2d 590, reversing (N.D. GA 1981) 510 Fed. Supp 668.

[16] 2 Cal. 4th 1, 825 P. 2d 388, 5 Cal. Rptr. 2d 495.

Do I Need to Take a Sample?

A friend of our family told me a moving story recently about the days before her husband's death. "Our health insurance wouldn't cover the costs of his treatment for the final four years because the drug he was using wasn't approved."

"Four years is a long time," I replied, thinking with my head and not my heart, "you'd think the drug would have been approved after the first year or so, once they'd acumulated enough evidence with other patients."

She shook her head. "His drug was never approved; other patients had trouble with it apparently, but it helped my husband a lot."

I thought about the many things I might have said in rebuttal, about spontaneous remissions and the need for large controlled samples in which we examine the experiences of multiple patients under a variety of conditions, but I let it go. Words would not have eased her pain.

Her husband took a drug and felt better for four years. The two events might have been related or they might have been completely independent. Probably the latter, as the experiences of the vast majority of patients with this same drug had been negative.

The limited experience of one or two individuals can never constitute proof. Only when we examine a large representative sample can we begin to draw conclusions that extend to an entire population.

2.8. To Learn More

The tale of a lady tasting tea appears in Fisher [1935] along with some excellent advice on the design of experiments. The tale is revisited by Neyman [1950; pp272–294]. Avoid their advice at your peril; poor experimental design may lead to the so-called "spiral after-effect" and endless misinterpretation, Stilson [1966]. Insight into the relationship between sample and population can be obtained from the opening chapters of Fisher [1925], Kempthorne [1955] and Hill[1966]. Hardy and Youse [1984] provide a careful review of probability fundamentals along with many exercises.

2.9. Exercises

1. Characterize the following relationships as positive linear, negative linear, or nonlinear:

   a. Distance from the top of a bathtub to the surface of the water as a function of time when you fill the tub, then pull the plug.

   b. Same problem only now you turn on the taps full blast.

   c. Sales as a function of your advertising budget.

   d. Blood pressure as you increase the dose of a blood-pressure-lowering medicine.

e. Electricity use as a function of the day of the year.

f. Number of bacteria at the site of infection as you increase the dose of an antibiotic.

g. Size of an untreated tumor over time.

2. Construct a scatter plot for the law school data in Exercise 1.14, if you haven't already done so. What do you think is the nature of the association between undergraduate GPA and LSAT score?

3. Which is the cause and which is the effect?

   a. Overpopulation and poverty.

   b. Highway speed limits and number of accidents.

   c. Cases of typhus and water pollution.

4. The Nielsen organization conducts weekly surveys of television viewing in the United States and issues estimates of the viewing audience for hundreds of TV programs.

   a. What is the population from which the samples are drawn?

   b. What would be an appropriate sampling method?

   c. What kinds of decisions are made based on these studies?

5. Suppose you have a six-sided die. What is the probability of rolling a 6? What is the probability of rolling a 6 twice in a row? What is the probability of rolling a pair of dice so the sum of their spots is 7? (Hint: Add up the probabilities of the separate mutually exclusive outcomes.)

6. Are you psychic? Place a partition between you and a friend so that you cannot see through the partition. Have your friend roll a die three times. Each time your friend rolls the die, write down what you think the result was. Compare your guesses with what was actually observed. Would you be impressed if you got one out of three correct? Two out of three?

7. An endocrinologist writing in the November 1973 issue of the *Ladies Home Journal* reported that "I take strong exception with the prevailing medical opinion that most "tired" women are neurotic, bored, or sex-starved ... . In my opinion, at least half such women can be shown to have a definite physical problem that causes their chronic illness." Are you convinced? Why or why not?

8. A study of 300 households in Santa Monica showed that a household produced an average of 4.5 pounds of garbage each day (0.8 pounds of recyclable bottles, 1.0 pound of yard clippings, 0.7 pounds of paper, and 2 pounds of yucky stuff).

   a. What is the sample in this study?

   b. How many observations are in the sample?

   c. What are the variables?

   d. Do the results of this study sound as if they would be typical of your household?

   e. What is the population from which this sample is drawn?

f. Would you be willing to extend inferences from this sample to this population?

9. Do the following constitute independent observations?

   a. Number of abnormalities in each of several tissue sections taken from the same individual.
   b. Sales figures at Eaton's department store for its lamp and cosmetic departments.
   c. Sales figures at Eaton's department store for the months of May through November.
   d. Sales figures for the month of August at Eaton's department store and its chief competitor Simpson-Sears.
   e. Opinions of several individuals whose names you obtained by sticking a pin through a phone book, and calling the "pinned" name on each page.
   f. Dow Jones index and GNP of the United States.
   g. Today's price in Australian dollars of the German mark and the Japanese yen.

10. Construct a theory to explain the opinion data recorded in Exercise 1.3. How would you go about selecting a sample to test your theory? Would you ask the same or different questions as those in the exercise?

11. Reexamine the executive compensation data of Exercise 1.2. Do you feel there is a relationship between executive compensation and sales? Executive compensation and return on equity? Can you encapsulate your gut feelings in model form?

12. Do you remember your own growth history? (I bet your mother does.) Plot a growth curve for yourself (your child, your grandchild) similar to that of Figure 2.3.

13. The values reported in Exercise 1.16 all relate to the same medical procedure. How would you explain the wide differences among them? How would you go about comparing the billing practices of the four Swedish hospitals?

14. Suppose you wanted to predict the sales of automobile parts; which observations would be critical? Number of automobiles in service? Number of automobiles under warranty? Number of automobiles with more than a given number of years in service? M2, the total money supply that is readily available in cash, savings and checking accounts, and so on? Retail sales. Consumer confidence. Weather forecasts. Would your answer depend on whether you owned a dealership or an independent automobile parts chain? (By the way, in the studies Mark Kaiser and I completed, the weather was the most significant variable; go figure.)

15. Draw the cumulative distributions of two sets of observations where the median of one set of observations is larger than the other.

CHAPTER 3

Testing Hypotheses

"Plan, Do, Check" is the strategy of big business in the 1990's; it's always been the underlying policy in science. In Chapter 1, we learned how to describe a sample and how to use the sample to describe and estimate the parameters of the population from which it was drawn. In Chapter 2, we learned how to develop hypotheses and frame them in the form of quantitative models. In this chapter, you'll learn how to test the hypotheses and models you've developed.

3.1. Two-Sample Comparison

3.1.1. "I Lost the Labels"

Shortly after I received my doctorate in statistics[1], I decided if I really wanted to help bench scientists apply statistics I should become a scientist myself. So back to school[2] I went to learn all about physiology and aging in cells raised in petri dishes.

I soon learned there was much more to an experiment than the random assignment of subjects to treatments. In general, 90% of the effort was spent mastering various arcane laboratory techniques, 9% developing new techniques to span the gap between what had been done and what I really wanted to do, and a mere 1% on the experiment itself. But the moment of truth came finally—it had to if I were to publish and not perish—and I succeeded in cloning human diploid fibroblasts in eight culture dishes. Four of these dishes were filled with a conventional nutrient

---

[1] From the University of California at Berkeley.
[2] The Wistar Institute, Philadelphia, PA, and the W. Alton Jones Cell Science Center in Lake Placid, NY.

solution and four held an experimental "life-extending" solution to which Vitamin E had been added.

I waited three weeks with my fingers crossed—there is always a risk of contamination with cell cultures—but at the end of this test period three dishes of each type had survived. My technician and I transplanted the cells, let them grow for 24 hours in contact with a radioactive label, and then fixed and stained them before covering them with a photographic emulsion.

Ten days passed and we were ready to examine the autoradiographs. Two years had elapsed since I first envisioned this experiment and now the results were in: I had the six numbers I needed.

"I've lost the labels," my technician said as he handed me the results.

"What!?" Without the labels, I had no way of knowing which cell cultures had been treated with Vitamin E and which had not.

"121, 118, 110, 34, 12, 22." I read and reread these six numbers over and over again. If the first three counts were from treated colonies and the last three were from untreated, then I had found the fountain of youth. Otherwise, I really had nothing to report.

3.2. Five Steps to a Permutation Test

How did I reach this conclusion?

In this and succeeding chapters, you will apply permutation techniques to a wide variety of testing problems ranging from simple to complex. In each case, you will follow the same five-step procedure we follow in this example:

1. Analyze the problem; identify the hypothesis and the *alternative (s)* of interest.
2. Choose a test statistic
3. Compute the test statistic for the original labeling of the observations.
4. Resample the labels and recompute the test statistic for the rearranged labels. Repeat until you obtain the distribution of the test statistic for all possible rearrangements.
5. Accept or reject the hypothesis using this permutation distribution as a guide.

3.2.1. *Analyze the Problem*

Let's take a more formal look at the problem of the missing labels. First, we'll identify the *hypothesis* and the *alternative (s)* of interest.

I wanted to assess the life-extending properties of a new experimental treatment with Vitamin E. To do this, I divided my cell cultures into two groups: one grown in a standard medium and one grown in a medium containing vitamin E. At the conclusion of the experiment and after the elimination of several contaminated cultures, both groups consisted of three independently treated dishes.

My null hypothesis was that the growth potential of a culture would not be affected by the presence of Vitamin E in the media; all the cultures would have

---

Five Steps to a Permutation Test

1. Analyze the problem:

    a. What is the hypothesis? What are the alternatives?
    b. How is the data distributed?
    c. What losses are associated with bad decisions?

2. Choose a statistic that will distinguish the hypothesis from the alternative.
3. Compute the test statistic for the original observations.
4. Rearrange the observations:

    a. Compute the test statistic for the new arrangement
    b. Compare the new value of test statistic with the value obtained for the original observations.
    c. Repeat steps a) and b) until you are ready to make a decision.

5. Draw a conclusion. Reject the hypothesis and accept the alternative if the value of the test statistic for the observations as they were labeled originally is an extreme value in the permutation distribution of the statistic. Otherwise, accept the hypothesis and reject the alternative.

---

equal growth potential. The alternative of interest was that cells grown in the presence of Vitamin E would be capable of many more cell divisions.

Under the null hypothesis, the labels "treated" and "untreated" provide no information about the outcomes, as the observations are expected to have more or less the same values in each experimental group differing as a result of some uncontrollable random fluctuation. If this hypothesis is true, we are free to exchange the labels.

3.2.2. Choose a Test Statistic

The next step in the permutation method is to choose a test statistic that discriminates between the hypothesis and the alternative. The statistic I chose was the sum of the counts in the group treated with Vitamin E. If the alternative is true, this sum should to be larger than the sum of the counts in the untreated group. If the null hypothesis is true, that is, if it doesn't matter which treatment the cells receive, then the sums of the two groups of observations should be approximately the same. One sum might be smaller or larger than the other by chance, but the two shouldn't be all that different.

3.2.3. Compute the Test Statistic

The third step in the permutation method is to compute the test statistic for each possible relabeling. But to compute the test statistic for the data as it had been labeled originally, I had to find the labels! Fortunately, I'd kept a record of the treatments independent of my technician. In fact, I'd deliberately not let my technician know which cultures were which in order to ensure she would give them equal care in handling. As it happened, the first three observations my technician showed me—121, 118, and 110—were those belonging to the cultures that received Vitamin E. The value of the test statistic for the observations as originally labeled is $349 = 121 + 118 + 110$.

3.2.4. Rearrange the Observations

We now rearrange, or *permute*, the observations, randomly reassigning the six labels, three "treated" and three "untreated," to the six observations. For example, treated, 121, 118, 34; untreated, 110, 12, 22. In this rearrangement, the sum of the observations in the first (treated) group is 273. We repeat this step until all $\binom{6}{3} = 20$ distinct rearrangements have been examined.[3]

	First Group			Second Group			Sum
1.	121	118	110	34	22	12	349
2.	121	118	34	110	22	12	273
3.	121	110	34	118	22	12	265
4.	118	110	34	121	22	12	262
5.	121	118	22	110	34	12	261
6.	121	110	22	118	34	12	253
7.	121	118	12	110	34	22	251
8.	118	110	22	121	34	12	250
9.	121	110	12	118	34	22	243
10.	118	110	12	121	34	22	240
11.	121	34	22	118	110	12	177
12.	118	34	22	121	110	12	174
13.	121	34	12	118	110	22	167
14.	110	34	22	121	118	12	166
15.	118	34	12	121	110	22	164
16.	110	34	12	121	118	22	156
17.	121	22	12	118	110	34	155
18.	118	22	12	121	110	34	152
19.	110	22	12	121	118	34	144
20.	34	22	12	121	118	110	68

3.2.5. Draw a Conclusion

The sum of the observations in the original Vitamin E treated group, 349, is equaled only once and never exceeded in the twenty distinct random relabelings. If chance alone is operating, such an extreme value is a rare, only-one-time-in-twenty event. I reject the null hypothesis at the 5% (1-in-20) significance level and embrace the alternative that the treatment is effective and responsible for the observed difference.

---

[3] The determination of the number of rearrangements, "6 choose 3" here, is considered in Section 2.2.2.1

In using this decision procedure, I risk making an error and rejecting a true hypothesis one in every twenty times. In this case, I did make just such an error. I was never able to replicate the observed life-promoting properties of Vitamin E in other repetitions of this experiment. Good statistical methods can reduce and contain the probability of making a bad decision, but they cannot eliminate the possibility.

Group 1	Group 2	Sum
1 2 3 4 10	5 6 7 8 9	20
1 2 3 4 5	6 7 8 9 10	15
1 2 3 4 6	5 7 8 9 10	16
1 2 3 4 7	5 6 8 9 10	17
1 2 3 4 8	5 6 7 9 10	18
1 2 3 4 9	5 6 7 8 10	19
1 2 3 5 6	4 6 7 8 10	17
1 2 3 5 7	4 6 8 9 10	18
1 2 3 5 8	4 6 7 9 10	19
1 2 3 5 9	4 6 7 8 10	20
and so forth		

3.3. A Second Example

Suppose we have two samples. The first, control, sample takes values 0, 1, 2, 3, and 19. The second, treatment, sample takes values 3.1, 3.5, 4, 5, and 6. Does the treatment have an effect?

In technical terms, we want to test that the median of the control population (or the mean, or the population distribution itself) is the same as the median (mean, distribution) of the vitamin E treated population, against the alternative that the median (mean) is higher or the frequency distribution is shifted to the right.[4]

The answer would be immediate if it were not for the value 19 in the first sample. The presence of this extreme value increases the mean of the first sample from 1.5 to 5. To dilute the effect of this extreme value on the results, we convert all the data to ranks, giving the smallest observation a rank of 1, the next smallest the rank of 2, and so forth. The first sample includes the ranks 1, 2, 3, 4, and 10 and the second sample includes the ranks 5, 6, 7, 8, 9. Is the second sample drawn from a different population than the first?

Let's count. The sum of the ranks in the first sample is 20. All the rearrangements with first samples of the form 1, 2, 3, 4, k, where k is chosen from the values {5, 6, 7, 8, 9, or 10} have sums that are as small as or smaller than that of our original sample. That's a total of six rearrangements. The four rearrangements whose first sample contains 1, 2, 3, 5, and a fifth number chosen from the set {6, 7, 8, 9} also have sums that are as small or smaller. That's $6 + 4 = 10$ rearrangements so far. Continuing in this fashion—I leave the complete enumeration for you as an

---

[4]See Exercise 14.

exercise—we find 13 of the $\binom{10}{5} = 252$ possible rearrangements have sums that are as small as or smaller than that of our original sample. Two samples this different will be drawn from the same population just over 5% of the time by chance.

3.3.1. More General Hypotheses

Our permutation technique is applicable whenever we can freely exchange labels; this is obviously the case when we're testing a null hypothesis. But suppose our hypothesis is somewhat different, that we believe, for example, that a new gasoline additive will increase mileage by at least twenty miles per tank. (Otherwise, why pay the big bucks for the additive?) Before using the additive, we recorded 295, 320, 329, and 315 miles per tank under various traffic conditions. With the additive, we recorded 330, 310, 345, and 340 miles per tank.

To perform the analysis, we begin by subtracting 20 from each of the latter figures. Our revised hypothesis is that the transformed mileages will be approximately the same, versus the alternative that the post additive transformed mileages will be less than the others. Our two transformed samples are

$$
\begin{array}{cccc}
295 & 320 & 329 & 315 \\
305 & 290 & 325 & 320
\end{array}
\tag{3.1}
$$

The actual analysis is left to you as an exercise.

Two-Sample Test for Location

Hypothesis H: mean/medians of groups differ by d_o
Alternative K: mean/medians of groups differ by $d > d_o$

Assumptions:

1. Under the hypothesis, observations are exchangeable.[5]
2. All observations in the first sample come from the same distribution F.
3. All observations in the second sample come from the distribution G where $G(x) = F(x - d)$.

Procedure:

1. Transform the observations in the first sample by subtracting d_0 from each.
2. Compute the sum of the observations in the smallest sample.
3. Compute this sum for all possible rearrangements of the combined sample.

Draw a conclusion. Reject the hypothesis in favor of the alternative if the original sum is an extreme value.

---

[5] Independent identically distributed observations are exchangeable. In this example, the observations become exchangeable after we transform them by subtracting δ_0 from each observation in the first sample.

3.4. Comparing Variances

Precision is essential in a manufacturing process. Items that are too far out of tolerance must be discarded and an entire production line brought to a halt if too many items exceed (or fall below) designated specifications. With some testing equipment, such as that used in hospitals, precision can be more important than accuracy. Accuracy can always be achieved through the use of standards with known values, while a lack of precision may render an entire sequence of tests invalid.

There is no shortage of methods to test the hypothesis that two samples come from populations with the same inherent variability, but few can be relied on. Many methods promise an error rate (significance level) of 5% but in reality make errors as frequently as 8% to 20% of the time. Other methods for comparing variances have severe restrictions.

For example, a permutation test based on the ratio of the sample variances is appropriate only if the means of the two populations are the same or we know their values.

Good (1994) showed we can eliminate the effects of the unknown locations if we subtract the sample median from each observation. Suppose the first sample contains the observations $x_1, \cdots x_n$ whose median is $mdn\{x_j\}$; we begin by forming the deviates $x'_j = |x_j - mdn\{x_j\}|$ for $j = 1, \cdots n$.[6] Similarly, we form the set of deviates $\{y'_j\}$ using the observations in the second sample and their median.

If there is an odd number of observations in the sample, then one of these deviates must be zero. We can't get any information out of a zero, so we throw it away. If there is an even number of observations in the sample, then two of these deviates must be equal. We can't get any information out of the second one that we didn't get from the first, so we throw it away.

Good's test statistic S is the sum of the remaining deviations in the first sample, that is, $S = \sum_{j=1}^{n-1} x'_j$. We obtain its permutation distribution and the cut-off point for our test by considering all possible rearrangements of the deviations that remain in both the first and second samples.

To illustrate the application of this statistic, suppose the first sample consists of the measurements 121, 123, 126, 128.5, and 129 and the second sample consists of the measurements 153, 154, 155, 156, and 158. Thus $x'_1 = 5$, $x'_2 = 3$, $x'_3 = 2.5$, $x'_4 = 3$, and $S_o = 13.5$. And $y'_1 = 2$, $y'_2 = 1$, $y'_3 = 1$, and $y'_4 = 3$. There are $\binom{8}{4} = 70$ arrangements in all, of which only three yield values of the test statistic as extreme as or more extreme than our original value. We conclude there is no significant difference between the dispersions of the two manufacturing processes.[7]

Good's test requires that we be able to exchange the labels. We can exchange them if the population variances are equal (as they would be under the null hypoth-

---

[6] Alternately, we might use the formula $x'_j = |x_j - mdn\{x_j\}|^2$.

[7] With such a small number of rearrangements, the evidence is unconvincing. We should continue to take observations until the evidence is overwhelming in one direction or the other.

esis), and if the only other difference between the two populations from which the samples are drawn is that they have different means.

The statistic δ proposed by Aly [1990][8] doesn't even have this latter restriction.

$$\delta = \sum_{i=1}^{m-1}(i)(m-i)(X_{(i+1)} - X_{(i)})$$

where $X_{(1)} < X_{(2)} < \cdots < X_{(m)}$ are the order statistics of the first sample.

Applying this statistic to the samples considered earlier, $X_{(1)} = 121, X_{(2)} = 123$, and so forth.

Set $z_{1i} = X_{(i+1)} - X_{(i)}$ for $i = 1, \ldots 4$. In this instance, $z_{11} = 123 - 121 = 2$, $z_{12} = 3$, $z_{13} = 2.5$, and $z_{14} = 0.5$.

The original value of the test statistic is $8 + 18 + 7.5 + 2 = 35.5(8 = 1 * 4 * 2$, and so forth). To compute the test statistic for other arrangements, we also need to know the differences $z_{2i} = Y_{(i+1)} - Y_{(i)}$ for $i = 1, \ldots 4$. $z_{21} = 1, z_{22} = 1, z_{23} = 1$, and $z_{24} = 2$.

One possible rearrangement is $\{2, 1, 1, 2\}$, which yields a value of $S = 8 + 6 + 6 + 8 = 28$.

There are $2^4 = 16$ rearrangements in all, of which one $\{2, 3, 2.5, 2\}$ is more extreme than the original observations. With 2 out of 16 rearrangements as extreme as or more extreme than the original, we accept the null hypothesis.[9]

If our second sample is larger than the first, we have to resample in two stages. First, we select a subset of m values $\{Y*_i, i = 1, \ldots m\}$ without replacement from the n observations in the second sample, and we compute the order statistics $Y*_{(1)} < Y*_{(2)} < \cdots < Y*_{(m)}$ and their differences $\{z_{2i}\}$. Last, we examine all possible values of Aly's measure of dispersion for permutations of the combined sample $\{\{z_{1i}\}, \{z_{2i}\}\}$ as we did when the two samples were equal in size and compare Aly's measure for the original observations with this distribution.

A similar procedure, first suggested by Baker [1995], should be used with Good's test in the case of unequal samples.

3.5. Pitman Correlation

We can easily generalize from two samples to three or more if the samples are ordered in some way, by the dose administered, for example.

Frank, Trzos, and Good [1978] studied the increase in chromosome abnormalities and micronucleii as the dose of various known mutagens was increased. Their object was to develop an inexpensive but sensitive biochemical test for mutagenicity that would be able to detect even marginal effects. The results of their experiment are reproduced in Table 3.1.

---

[8]The choice of test statistic is dictated by the principles of density estimation discussed in Chapter 10.

[9]Again, the wiser course would be to take a few more observations.

TABLE 3.1. Micronucleii in Polychromatophilic Erythrocytes and Chromosome Alterations in the Bone Marrow of Mice Treated with CY

Dose (mg/kg)	Number Animals	Micronucleii per 200 Cells	Breaks per 25 Cells
0	4	0 0 0 0	0 1 1 2
5	5	1 1 1 4 5	0 1 2 3 5
20	4	0 0 0 4	3 5 7 7
80	5	2 3 5 11 20	6 7 8 9 9

To analyze such data, Pitman [1937] proposes a test for linear correlation with three or more ordered samples using as test statistic $S = \Sigma g[i]x_i$, where x_i is the sum of the observations in the ith dose group and $g[i]$ is any monotone increasing function.[10] The simplest example is $g[i] = i$, with test statistic $S = \Sigma i x_i$. In this instance, we take $g[\text{dose}] = \log[\text{dose} + 1]$, as the anticipated effect is proportional to the logarithm of the dose.[11] Our test statistic is $S = \Sigma \log[\text{dose}_i + 1]x_i$.

The original data for breaks may be written in the form

$$0112 \quad 01235 \quad 3577 \quad 67899$$

As $\log[0 + 1] = 0$, the value of the Pitman statistic for the original data is $0 + 11 * \log[6] + 22 * \log[21] + 39 * \log[81] = 112.1$. The only larger values are associated with the small handful of rearrangements of the form

$$
\begin{array}{llll}
0012 & 11235 & 3577 & 67899 \\
0011 & 12235 & 3577 & 67899 \\
0011 & 12233 & 5577 & 67899 \\
0012 & 11233 & 5577 & 67899 \\
0112 & 01233 & 5577 & 67899 \\
0112 & 01235 & 3567 & 77899 \\
0012 & 11235 & 3567 & 77899 \\
0011 & 12235 & 3567 & 77899 \\
0011 & 12233 & 5567 & 77899 \\
0012 & 11233 & 5567 & 77899 \\
0112 & 01233 & 5567 & 77899 \\
\end{array}
$$

Because there are $\binom{18}{4\,5\,4}$ rearrangements in all,[12] a statistically significant ordered dose response ($\alpha < 0.001$) has been detected. The micronucleii also exhibit a

---

[10] A monotone increasing function $g[x]$ increases steadily as x gets larger. The simplest example is $g[i] = i$. Another is $g[i] = ai + b$, with $a > 0$. The permutation distributions of S_1 with $g[i] = ai + b$ and S with $f[i] = i$ are equivalent in the sense that if S^o, S_1^o are the values of these test statistics corresponding to the same set of observations $\{x_i\}$, then $\Pr(S > S^o) = \Pr(S_1 > S_1^o)$.

[11] Adding a 1 to the dose keeps this function from blowing up at a dose of zero.

[12] $\binom{18}{4\,5\,4} = 18!/(4!5!4![18 - 4 - 5 - 4]!)$, the number of ways 18 things can be divided among four subgroups of sizes 4, 5, 4, and 5.

---

k-Sample Test for Ordered Samples

Hypothesis H: All distributions and, in particular, all population means, are the same.

Alternative K: The population means are ordered.

Assumptions:

1. Observations are exchangeable.
2. All the observations in the ith sample come from the same distribution G_i, where $G_i(x) = F(x - \delta_i)$.

Test statistic:

$S = \Sigma g[i] n_i x_i$, where n_i is the number of observations in the ith sample.

---

statistically significantly dose response when we calculate the permutation distribution of $S = \Sigma \log[\text{dose}_i + 1] n_i$. To perform the calculations, we took advantage of the C+ computer program detailed in Appendix 1; the only change was in the subroutine used to compute the test statistic.

A word of caution: If we use some function of the dose other than $g[\text{dose}] = \log[\text{dose} + 1]$, we might observe a different result. Our choice of a test statistic must always make practical as well as statistical sense.

3.5.1. Effect of Ties

Ties can complicate the determination of the significance level. Because of ties, each rearrangement noted in the preceding example might actually have resulted from several distinct reassignments of subjects to treatment groups and must be weighted accordingly. To illustrate this point, suppose we put tags on the 1s in the original sample:

$$0 \; 1 * \; 1 \# \; 2 \quad 0 \; 1 \; 2 \; 3 \; 5 \quad 3 \; 5 \; 7 \; 7 \quad 6 \; 7 \; 8 \; 9 \; 9$$

The rearrangement

$$0 \; 0 \; 1 \; 2 \quad 1 \; 1 \; 2 \; 3 \; 5 \quad 3 \; 5 \; 7 \; 7 \quad 6 \; 7 \; 8 \; 9 \; 9$$

corresponds to the three reassignments

$$0 \; 0 \; 1 \quad 2 \quad 1 * \; 1 \# \; 2 \; 3 \; 5 \quad 3 \; 5 \; 7 \; 7 \quad 6 \; 7 \; 8 \; 9 \; 9$$
$$0 \; 0 \; 1 * \; 2 \quad 1 \quad 1 \# \; 2 \; 3 \; 5 \quad 3 \; 5 \; 7 \; 7 \quad 6 \; 7 \; 8 \; 9 \; 9$$
$$0 \; 0 \; 1 \# \; 2 \quad 1 \quad 1 * \; 2 \; 3 \; 5 \quad 3 \; 5 \; 7 \; 7 \quad 6 \; 7 \; 8 \; 9 \; 9$$

The 18 observations are divided into four dose groups containing 4, 5, 4, and 5 observations, respectively, so that there are $\binom{18}{4\,5\,4}$ possible reassignments of observations to dose groups. Each reassignment has probability $1/\binom{18}{4\,5\,4}$ of occurring so the probability of the rearrangement

$$0 \; 0 \; 1 \; 2 \quad 1 \; 1 \; 2 \; 3 \; 5 \quad 3 \; 5 \; 7 \; 7 \quad 6 \; 7 \; 8 \; 9 \; 9$$

Test for Bivariate Dependence

H: The observations within each pair are independent of one another.
K: The observations within each pair are all positively correlated or all negatively correlated.

Assumptions:

1. (x_i, y_i) denotes the members of the ith pair of observations.
2. The $\{x_i\}$ are exchangeable and the $\{y_i\}$ are exchangeable.

Test statistic:
$S = \Sigma y_i x_i$: Reject hypothesis of independence if the value of S for the observations before they are rearranged is an extreme value in the permutation distribution.

---

is $3\binom{18}{4\ 5\ 4}$.

To determine the significance level when there are ties, weight each distinct rearrangement by its probability of occurrence. This weighting is done automatically if you use Monte Carlo sampling methods.[13]

3.6. Bivariate Dependence

Are two variables correlated with one another? Is there some sort of causal relationship between them or between them and a third hidden variable? Take another look at Figure 1.4 in Chapter 1. Surely arm span and height must be closely related. We can use the Pitman correlation method introduced in the previous section to verify our conjecture. If $(a_1, h_1), (a_2, h_2) \cdots (a_n, h_n)$ is a set of n observations on arm span and height, then to test for association between them we need to look at the permutation distribution of the Pitman correlation

$$S = \sum_{i=1}^{n} a_i h_i.$$

To illustrate this approach, let's look at the arm span and height of the five shortest students in my sixth-grade class:

$$(139, 137) \quad (140, 138.5) \quad (141, 140) \quad (142.5, 141) \quad (143.5, 142)$$

Both the arm spans and the heights are in increasing order. Is this just coincidence? What is the probability that an event like this could happen by chance alone? We

---

[13] In a Monte Carlo simulation, we let the computer choose a random subset of the possible rearrangements and use this subset to estimate the probability of finding a value of the test statisic as extreme as or more extreme than that which was actually observed.

could list all possible permutations of both arm span and height, but this isn't necessary. We can get exactly the same result if we fix the order of one of the variables, the height, for example, and look at the 5! = 120 ways in which we could rearrange the arm span readings:

(140, 137)   (139, 138.5)   (141, 140)   (142.5, 141)   (143.5, 142)
(141, 137)   (140, 138.5)   (139, 140)   (142.5, 141)   (143.5, 142)

and so forth.

Obviously, the arrangement we started with is the most extreme, occurring exactly 1 time in 120 by chance alone. Applying this same test to all 22 pairs of observations, we find the odds are less than 1 in a million that what we observed occurred by chance alone and we conclude that arm span and height are directly related.

3.7.   One-Sample Tests

The resampling methods that work for two, three, or more samples, won't work for one. Another approach is required. In this section, we consider two alternate approaches to the problem of testing a hypothesis concerning a single population—the bootstrap and the permutation test.

3.7.1.   The Bootstrap

Suppose we wish to test the hypothesis that the mean temperature of a process is 440°C and we have already taken 20 readings (see Table 3.2). First, let us apply the bootstrap, which was introduced in Section 1.5.4. We need only assume that the 20 readings are independent. The mean reading is 454.6. This represents a deviation from our hypothesized mean temperature of 14.6 degrees. How likely is a deviation this large to arise by chance? To find out, we draw 100 bootstrap resamples. The results are depicted in Table 3.3 and Figure 3.1. In not one of these 100 bootstrap samples did the mean fall below 444.95. We reject the hypothesis that the mean of our process is 440°C.

TABLE 3.2. Process Temperature (from Cox and Snell, 1981)

431, 450, 431, 453, 481, 449, 441, 476, 460, 482, 472, 465,
421, 452, 451, 430, 458, 446, 466, 476.

TABLE 3.3. Percentiles of the Resampling Distribution

1%	445.15	5%	447.8	10%	449.42
25%	452.1	50%	454.3	75%	456.6
90%	458.98	95%	459.52	99%	462.7

FIGURE 3.1. Scatterplot of Bootstrap Median Temperatures

One caveat: This test is not an exact one; the actual significance level may be greater than 1%. (Tips on reducing the level error will be found in Chapter 5.)

3.7.2.   Permutation Test

We can achieve an exact significance level using the permutation approach if we can make two assumptions: (i) the readings in Table 3.3 are independent; and (ii) they come from a symmetric distribution.

First, we create a table of the deviations about 440, the hypothesized process mean (Table 3.4). Right off the bat, we notice something suspicious. If the underlying distribution is symmetric so that positive and negative deviations are equally likely, why are only 4 of the 20 deviations negative? This could happen less than 1 in 100 times by chance (see Section 2.2.2.2), and we should reject our hypothesis that the process mean is 440.

Our test is a sign test based on the binomial, rather than a permutation test, but why waste time when you don't need to? The sign test is the optimal approach when observations are dirt cheap. When data is expensive or difficult to obtain, we may need a more powerful test. To more fully illustrate the permutation test for a single sample, let's consider a test of a hypothesized mean value of 450, again using the data of Table 3.2.

Now the decision isn't so obvious. Examining Table 3.5, we see that 7 of the 20 deviations are negative, and there are large negative deviations as well as large positive ones.

Suppose we lost track of the signs (the way we lost track of the labels in my earlier example). Because the underlying distribution is symmetric—our fundamental assumption—we can attach new signs at random, selecting a plus or minus with equal probability, for a total of $2 \times 1 \times 2 \times 2 \times \cdots$ or 2^{19} possible rearrangements.

TABLE 3.4. Process Temperature Deviations from a Hypothesized Mean Value of 440

431	+9	476	+36	451	+11
450	+10	460	+20	430	−10
431	−9	482	+42	458	+18
453	+13	472	+32	446	+6
481	+41	465	+25	466	+26
449	+9	421	−19	476	+36
441	+1	452	+12		

TABLE 3.5. Process Temperature Deviations from a Hypothesized Mean Value of 450

431	−19	476	+26	451	+1
450	+0	460	+10	430	−20
431	−19	482	+32	458	+8
453	+3	472	+22	446	−4
481	+31	465	+15	466	+16
449	−1	421	−29	476	+26
441	−9	452	+2		

Estimating Permutation Test Significance Level

Get the data and pack it into a single linear vector.
Compute the test statistic for the original samples.
Repeat the following several hundred times:
  Resample from the data:
    Select new samples from the combined data set.
    Select at random with replacement.
  Compute test statistic for the rearrangement.
  Compare new value of test statistic with original value.
Estimate significance level.

Estimating Bootstrap Significance Levels

Get the data.
Compute the test statistic for the original samples.
Repeat the following several hundred times:
  Resample from the data:
    Select separately from each of the original samples.
    Select at random with replacement.
  Compute test statistic for the rearrangement.
  Compare new value of test statistic with original value.
Estimate significance level.

(The 0 doesn't get rearranged.) The sum of the negative differences for our current rearrangement is −101. Alas, there are far too many rearrangements, 2^{20} or one million million, for us to even attempt a count. A random sample of 10,000 rearrangements yields an estimate of the probability of a sum of negative differences as small as that observed in Table 3.5 on the order of 13%. We accept the hypothesis that the mean is 450.

3.7.3. Automating the Process

Tabulating by hand is feasible only for very small samples with a correspondingly small number of rearrangements. Consider that while $5! = 120$, and $10! = 463,600$. For larger samples, a computer is required. A few commercially available statistical packages perform a wide variety of permutation tests and provide exact results—see Appendix 4. Others can be programmed to do the work—see sidebar and Appendixes 1, 2, and 3. In Stata, for example, simply type **sample 100** to generate a single random rearrangement. Of course, you'll have to write additional code to compute the test statistic, cumulate, and generate repeated rearrangements. Even with a computer, we may be forced to be content with a random sample of the possibilities. Those 2^{20} rearrangements in the preceding example would tie up our computer for years.

3.8. Transformations

The applicability of the permutation test in the one-sample case, matched pairs, and regression and of the bootstrap in all cases can be improved if a prior transformation is used that makes the underlying distribution symmetric.

If unwilling to give undo weight to exceptionally large deviations, a rank transformation should be done initially (see Section 3.2).

If working with an index or ratio where changes are measured in percent rather than absolute terms, for example, virus titres, a logarithmic transformation is recommended.

When working with observations that denote the number of times some event of interest has occurred, for example, the number of surviving patients or the number of one-legged Republicans, the arc sine transformation will often produce the desired result.[14]

3.9. Matched Pairs

We can immediately apply our results for a single sample to matched pairs. In a matched-pairs experiment, each subject in the treatment group is matched as closely as possible by a subject in the control group. For example, if a 45-year-old black male hypertensive is given a blood-pressure lowering pill, then we give a second similarly built 45-year-old black male hypertensive a placebo.[15] One

---

[14] Don't be concerned if you don't know what the arc sine transformation is, most statistics programs include it as one of the essential built-in functions.

[15] A placebo looks and tastes like a medicine but consists of an innocuous chemical or filler without any real biological effect. Does it work? Many patients get better with the help of a placebo, especially if a doctor, a witch doctor, a parent, or a significant other gives it to them.

member of each pair is then assigned at random to the treatment group, and the other member is assigned to the controls.

The value of this approach is that it narrows our focus to differences arising solely from treatment, reducing, through matching, the noise, or *dispersion*, resulting from differences among individuals.

Assuming we've been successful in our matching, we end up with a series of independent pairs of observations (X_i, Y_i) where the members of each pair have been drawn from distributions that have been shifted with respect to one another as in Figure 3.2. Regardless of the form of this unknown distribution, the differences $Z_i = X_i - Y_i$ will be symmetrically distributed about the unknown shift parameter.

3.9.1. An Example: Pre- and Post-Treatment Levels

In this example, pretreatment and post-treatment serum antigen levels were measured in 20 AIDS patients. The data was entered into StatXact^{TM} and a permutation test for two related samples used to test whether there is a statistically significant effect due to treatment (see Table 3.6).

```
Analysis with StatXact 3

Datafile: C:\SX3WIN\EXAMPLES\AZT1.CY3
```

TABLE 3.6. Response of Serum Antigen Level to AZT *Source*: Makutch and Parks [1988]. Serum Level (pg/ml).

Patient ID	Pre	Post	Difference
0 1	149	0	−149
0 2	0	51	51
0 3	0	0	0
0 4	259	385	126
0 5	106	0	−106
0 6	255	235	−20
0 7	0	0	0
0 8	52	0	−52
0 9	340	48	−392
0 10	0	65	65
0 11	180	77	−103
0 12	0	0	0
0 13	84	0	−84
0 14	89	0	−89
0 15	212	53	−159
0 16	554	150	−404
0 17	500	0	−500
0 18	424	165	−259
0 19	112	98	−14
0 20	2,600	0	−2, 600

PERMUTATION TEST FOR TWO RELATED SAMPLES

Summary of Exact distribution of PERMUTATION TEST statistic:

Min	Max	Mean	Std-dev	Observed	Standardized
0.0000	5108.	2554.	1369.	177.0	-1.736

Exact Inference:
      One-sided p-value: Pr{Test Statistic .LE. Observed} = 0.0011
      Pr{Test Statistic .EQ. Observed} = 0.0000
      Two-sided p-value:Pr{|Test Statistic-Mean|}
      .GE.|Observed-Mean| = 0.0021
      Two-sided p-value: 2 * One-Sided = 0.0021

A highly significant difference![16]

If we use Student's t in this example, as some textbooks recommend (see Section 4.3.2), we would obtain a one-sided p-value of 0.0413 and a two-sided p-value of 0.0826. Which test is correct? The significance levels derived from Student's t are exact when the underlying observations are independent and each is drawn from the same normal distribution. The permutation test is exact even if the underlying distribution is not normal, as long as the observations are independent and identically distributed.

3.10.   Summary

In this chapter you learned how to perform a permutation test along with specific tests for comparing the locations and dispersions of two populations. Pitman correlation, which you studied here, enables you to compare the locations of several populations against an ordered alternative and to test for bivariate dependence. You

---

Matched Pairs

Hypothesis H: Means/medians of the members of each pair are the same.
Alternative K: Means/medians of the members of each pair differ by $d > 0$.

Assumption:
Independent observations.

First, compute the differences $z_i = x_i - y_i$ for each pair of observations.
Compute the test statistic, the sum of the positive $z_i : S = \sum_{z_i > 0} z_i$.

---

[16] An effect that is statistically significant may not be biologically significant.

used two different resampling methods—the permutation test and the bootstrap—to test hypotheses regarding the location parameter of a single population. And you saw that you could use these same tests to compare two populations via matched pairs.

3.11. To Learn More

The permutation tests were introduced by Pitman [1937, 1938] and Fisher [1935] and have seen wide application in chemistry [vanKeerberghen et al 1991], clinical trials [Gail, Tan, and Piantadosi 1988], Howard 1981], cybernetics [Valdesperez 1995], ecology [Busby 1990; Cade 1997; Pollard, Lakhand and Rothery 1987; Prager and Hoenig 1989], law [Gastwirht 1992], education [Gliddentracy and Greenwood 1997; Gliddentracy and Parraga 1996], medicine [Feinstein 1973; Tsutakawa and Yang 1974], meteorology [Gabriel 1979; Tukey 1985], neurology [Ford, Colom, and Bland 1989], pharmacology [Plackett and Hewlett 1963], physiology [Boess et al 1990; Faris and Sainsbury 1990; Zempo et al 1996], psychology [Hollander and Penna 1988; Hollander and Sethuramann 1978; Kazdin 1980l May, Masson and Hunter 1990], theology [Witzum, Rips, and Rosenberg 1994], toxicology [Farrar and Crump 1988, 1991] and zoology [Adams and Anthony 1996; Jackson 1990]. Texts dealing with their application include Bradley [1968], Edgington [1995], Maritz [1996], Noreen [1989], Manly [1997], and Good [1994]. See, also, the recent review by Barbella, Deby, and Glandwehr [1990]. Early articles include Pearson [1937], Wald and Wolfowitz [1944]. The problem of outliers, that is extreme, questionable observations, is considered by Lambert [1985], Good [1994], and Maritz [1996].

The use of Monte Carlo simulations to approximate the permutation distribution was proposed by Dwass [1957], Hope [1968], and Marriott [1979]. An excellent discussion of how many simulations to use in permutation tests may be found in the StatXact 3 manual by C. Mehta and N. Patel (see Appendix 4). The optimum number of bootstrap replications is considered in Efron and Tibshirani [1993; p50–53].

3.12. Exercises

1. How was the analysis of the cell culture experiment described in Section 3.1 affected by the loss of two of the cultures due to contamination? Suppose these cultures had escaped contamination and given rise to the observations 90 and 95; what would be the results of a permutation analysis applied to the new, enlarged data set consisting of the following cell counts:

| Treated | 121 | 118 | 110 | 90 |
| Untreated | 95 | 34 | 22 | 12 |

2. In the preceding example, what would the result have been if you had used as your test statistic the difference between the sums of the first and second samples? The difference between their means? The sum of the squares of the observations in the first sample? The sum of their ranks?

3. Is my Acura getting less fuel efficient with each passing year? Use the data in Exercise 1.11 to test the null hypothesis against this alternative.

4. How many different subsamples of size 5 can be drawn without replacement from a sample of size 9? From a sample of size 10?

5. Cognitive dissonance theory suggests that when people experience conflicting motives or thoughts, they will try to reduce the dissonance by altering their perceptions. For example, college students were asked to perform a series of repetitive, boring tasks. They were then paid either $1 or $20 to tell the next student that the tasks were really interesting and fun. In private, they were asked to rate their own feelings about the tasks on a scale from 0 (dumb, dull, and duller) to 10 (exciting, better than sex).

   Those who received $20 for lying, assigned ratings of 3, 1, 2, 4, 0, 5, 4, 5, 1, and 3. Those who received only $1, appeared to rate the tasks more favorably: 4, 8, 6, 9, 3, 6, 7, 10, 4, and 8. Is the difference statistically significant? If you aren't a psychologist and would like to know what all this proves, see Festinger and Carlsmith [1959].

6. Babies can seem awfully dull for the first few weeks after birth. To us it appears that all they do is nurse and wet, nurse and wet. Yet in actuality, their brains are incredibly active. (Even as your brain, dulled by the need to wake up every few hours during the night to feed the kid, has gone temporarily on hold.) Your child is learning to see, to relate nervous impulses received from two disparate retina into concrete visual images. Consider the results of the following deceptively simple experiment in which an elaborate apparatus permits the experimenter to determine exactly where and how long an infant spends in examining a triangle. (See Salapotek and Kessen, 1966, for more details and an interpretation.)

Subject	Corners	Sides
Tim A.	10	7
Bill B.	16	10
Ken C.	23	24
Kath D.	23	18
Misha E.	19	15
Carl F.	16	18
Hillary G.	12	11
Holly H.	18	14

   a. Is there a statistically significant difference in viewing times between the sides and corners of the triangle?

   b. Did you (should you) use the same statistical test as you used to analyze the cognitive dissonance data?

7. Use Pitman correlation to test whether DDT residues have a deleterious effect on the thickness of a cormorant's egg shell:

DDT residue in yolk (ppm)	65	98	117	122	393
Thickness of shell (mm)	.52	.53	.49	.49	.37

8. The following vaginal virus titers were observed in mice by H. E. Renis of the Upjohn Company 144 hours after inoculation with Herpes virus type II (see Good, 1979, for complete details):

Saline controls	10000,	3000,	2600,	2400,	1500
Treated with antibiotic	9000,	1700,	1100,	360,	1

  a. Is this a one-sample, two-sample, k-sample, or matched pairs study?
  b. Does treatment have an effect?
  c. Most authorities would suggest using a logarithmic transformation before analyzing this data. Repeat your analysis after taking the logarithm of each of the observations. Is there any difference? Compare your results and interpretations with those of Good [1979].

9. You think you know baseball? Do home run hitters have the highest batting averages? Think about this hypothesis, then analyze the following experience based on a half season with the Braves:

Batting average	.252	.305	.299	.303	.285	.191	283	.272
Home runs	12	6	4	15	2	2	16	6
Bating average	.310	.266	.215	.211	.244	.320		
Home runs	8	10	0	3	6	7		

If you don't believe this result, then check it out for your favorite team.
10. In Exercise 2.10, you developed a theory based on the opinion data recorded in Exercise 1.3. Test that theory.
11. Take a third look at the law school data in Exercise 1.14. Is there a correlation between undergraduate GPA and performance on the LSAT?
12. To compare teaching methods, a group of school children was randomly assigned to two groups. The following are the test results:

Conventional	65	79	90	75	61	85	98	80	97	75
New	90	98	73	79	84	81	98	90	83	88

Are the two teaching methods equivalent in result?
13. To compare teaching methods, ten schoolchildren were first taught by conventional methods, tested, and then taught by an entirely new approach. The following are the test results:

Conventional	65	79	90	75	61	85	98	80	97	75
New	90	98	73	79	84	81	98	90	83	88

Are the two teaching methods equivalent in result?
How does this experiment differ from that described in Exercise 12?

14. Sketch a diagram to show that if $G(x + 1) = F(x)$ where F, G are two distribution functions, all of the following are true: i) $G(x) < F(x)$ for all x, ii) G is shifted to the right of F, and iii) the median of G is larger by 1 than the median of F.

15. You can't always test a hypothesis. A marketing manager would like to show that an intense media campaign just before Christmas has resulted in increased sales. Should he compare this year's sales with last year's? What would the null hypothesis be? And what would some of the alternatives be?

16. How would you test the hypothesis that the fuel additive described in Section 3.3.1 increases mileage by 10%?

CHAPTER 4

When the Distribution Is Known

One of the strengths of the hypothesis-testing procedures described in the preceding chapter is that you don't need to know anything about the underlying population(s) to apply them. But suppose you do know something about these populations, that you have full knowledge of the underlying processes that led to the observations in your sample(s), should you still use the same statistical tests? The answer is *no, not always*, particularly with very small or very large amounts of data. In this chapter, we'll consider several parametric approximations, including the binomial (which you were introduced to in Chapter 2), the Poisson, and the normal or Gaussian, along with several distributions derived from the latter that are of value in testing location and dispersion.

4.1. Binomial Distribution

Recall that a binomial frequency distribution, written $B(n, p)$ results from a series of n independent trials, each with a probability p of success and a probability $1 - p$ of failure. The *mathematical expectation or expected value* of the number of successes in a single trial is p; this is the proportion of successes we would observe in a very large number of repetitions of a single trial. The expected number of successes in n independent trials would be pn.[1]

---

[1] That is, if we were to perform N replications of n binomial trials, each with probability p of success, where N is a very large number, the expected proportion of successes would be np.

4.1.1. Properties of Independent Observations

The *expected value* of an observation is the mean of all the values we would observe if we were to make a very large number of independent observations. In a population consisting of N individuals, the expected value of an individual's height h is the population mean $\frac{1}{N} \sum_{k=1}^{N} h_k$. It's easy to see that as a sample grows larger, the sample mean $\frac{1}{N} \sum_{k=1}^{N} h_k$ gets closer to the population mean.

The *expected value* of the sum of two independent observations X and Y is equal to the expected value of X plus the expected value of Y.

If μ is the expected value of X, the variance of X is defined as the expected value of $(X - \mu)^2$. In a population consisting of N individuals, the variance of an individual's height is the population variance $\frac{1}{N} \sum_{k=1}^{N} (h_k - \mu)^2$. As a sample grows larger and larger, the sample variance $s^2 = \sum_{i=1}^{n} (x_i - \bar{x})^2 / (n - 1)$ gets closer to the population variance.

If X and Y are the values taken by independent observations, the variance of $X + Y$ is the sum of the variance of X and the variance of Y. The variance of $X - Y$ is also the sum of the variance of X and the variance of Y.

The variance of the sum of n independent observations is the sum of their variances. If the observations are identically distributed, each with variance σ^2, the sum of their variances is $n\sigma^2$. The variance of their mean $\bar{x}$ is σ^2/n for

$$(\bar{x} - \mu)^2 = \left(\frac{1}{N} \sum_{k=1}^{N} x_k - \mu \right)^2 = \left(\frac{1}{N} \sum_{k=1}^{N} (x_k - \mu) \right)^2 = \frac{1}{N^2} \left(\sum_{k=1}^{N} (x_k - \mu) \right)^2$$

Thus, the mean of 100 observations has 1/100th the variance of a single observation. In this example, the standard deviation of the mean, termed the *standard error*, would have 1/10th the standard deviation of a single observation. We'll use this latter property of the mean in Chapter 6 when we try to determine how large a sample should be.

Let's apply these definitions to determine the variance of the expected number of successes in n independent binomial trials. If the probability of success is p, we would expect a proportion p of the n trials to be successful; that's np trials. We have two types of deviates, a proportion pn corresponding to successes that take the value $(1 - p)^2$ and a proportion $(1 - p)n$ corresponding to failures that take the value $(0 - p)^2$. A little algebra (Exercise 2) shows that the variance is $np(1 - p)$.

4.1.2. Testing Hypotheses

Suppose we've flipped a coin in the air seven times, and six times it's come down heads. Do we have reason to suspect the coin is not fair, that is, that p, the probability of throwing a head, is greater than 1/2?

To answer this question, we need to look at the frequency distribution of the binomial observation X with n independent trials and probability p of success for each trial.

$$\Pr\{X = j\} = \binom{n}{j} p^j (1 - p)^{n-j} \text{ for } j = 0, 1, 2, \cdots n.$$

If $n = 7$ and $k = 6$, this probability is $7p^6(1 - p)$. For $p = 1/2$, this probability is $6/128 = 0.468 < 5\%$. If six heads out of seven tries seems extreme to us, seven heads out of seven would have seemed even more extreme. Adding the probability of this more extreme event to what we have already, we see the probability of throwing six or more heads in seven tries is $7/128 = 0.547 > 5\%$. Still, six or more heads out of seven tries does seem suspicious. If you were a mad scientist and you observed that six times out of seven your assistant Igor began to bay like a wolf when there was a full moon, wouldn't you get suspicious?

4.2. Poisson: Events Rare in Time and Space

The decay of a radioactive element, an appointment to the United States Supreme Court, a cavalry officer trampled by his horse, have in common that they are relatively rare but inevitable events. They are inevitable, that is, if there are enough atoms, enough seconds or years in the observation period, enough horses and momentarily careless men. Their frequency of occurrence has a Poisson distribution.

The number of events in a given interval has the Poisson distribution if it is the cumulative result of a large number of opportunities, each of which has only a small chance of occurring. The interval can be in space as well as time. For example, if we seed a small number of cells into a petri dish that is divided into a large number of squares, the distribution of cells per square follows the Poisson.[2]

If an observation X has a Poisson distribution such that we may expect an average of λ events per interval,

$$Pr\{X = k\} = \frac{\lambda^k e^{-\lambda}}{k!} \text{ for } k = 0, 1, 2, \cdots.$$

An interesting and useful property of such observations is that the sum of a Poisson with mathematical expectation λ_1 and a second independent Poisson with mathematical expectation λ_2 is a Poisson with mathematical expectation $\lambda_1 + \lambda_2$.

4.2.1. Applying the Poisson

John Ross of the Wistar Institute held that there were two approaches to biology: the analog and the digital. The *analog* was served by the scintallation counter: one ground up millions of cells then measured whatever was left behind in the stew; the *digital* was to be found in cloning experiments where any necessary measurements would be done on a cell-by-cell basis.

---

[2]The stars in the sky do not have a Poisson distribution, although it certainly looks this way. Stars occur in clusters, and these clusters, in turn, are distributed in super-clusters. The centers of these super-clusters do follow a Poisson distribution in space, and the stars in a cluster follow a Poisson distribution around the center almost as if someone had seeded the sky with stars and then watched the stars seed themselves in turn. See Neyman and Scott [1960].

John was a cloner and, later, as his student, so was I. We'd start out with ten million or more cells in a 10-milliliter flask and try to dilute them down to one cell per milliliter. We were usually successful in cutting down the numbers to ten thousand or so. Then came the hard part. We'd dilute the cells down a second time by a factor of 1:100 and hope we'd end up with 100 cells in the flask. Sometimes we did. Ninety percent of the time, we'd end up with between 90 and 110 cells, just as the binomial distribution predicted. But just because you cut a mixture in half (or a dozen, or a hundred parts) doesn't mean you're going to get equal numbers in each part. It means the probability of getting a particular cell is the same for all the parts. With large numbers of cells, things seem to even out. With small numbers, chance seems to predominate.

Things got worse when I went to seed the cells into culture dishes. These dishes, made of plastic, had a rectangular grid cut into their bottoms, so they were divided into approximately 100 equal-size squares. Dropping 100 cells into the dish meant an average of 1 cell per square. Unfortunately for cloning purposes, this average didn't mean much. Sometimes, 40% or more would contain two or more cells. It didn't take long to figure out why. Planted at random, the cells obey the Poisson distribution. An average of one cell per square means

$$\Pr\{\text{No cells per square}\} = 1 * e^{-1}/1 = 0.32$$

$$\Pr(\text{One cell per square}) = 1 * e^{-1}/1 = 0.32$$

$$\Pr(\text{Two or more cells per square}) = 1 - 0.32 - 0.32 = 0.34.$$

Two cells was one too many; a clone must begin with a single cell. I had to dilute the mixture a third time to ensure that the percentage of squares that included two or more cells was vanishingly small. Alas, the vast majority of squares were now empty; I was forced to spend hundreds of additional hours peering through the microscope looking for the few squares that did include a clone.

4.2.2.   Comparing Two Poissons

Suppose in designing a new nuclear submarine (or that unbuilt wonder, a nuclear spacecraft) that you become concerned about the amount of radioactive exposure that will be received by the crew. You conduct a test of two possible shielding materials. During ten minutes of exposure to a power plant using each material in turn as a shield, you record 14 counts with material A and only 4 with experimental material B. Can you conclude that B is safer than A?

The answer lies not with the Poisson but the binomial. If the materials are equal in their shielding capabilities, then each of the 18 recorded counts is as likely to be obtained through the first material as through the second. In other words, you would be observing a B(18, 1/2). The numeric answer is left as an exercise.

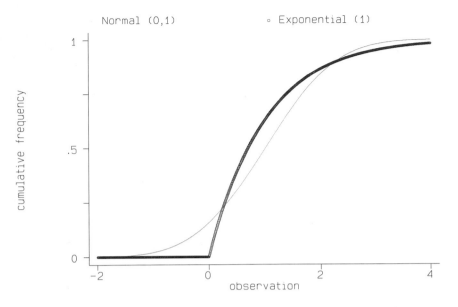

Normal (0,1)   ∘ Exponential (1)

cumulative frequency

FIGURE 4.1. Cumulative Distributions of Normal and Exponential Observations with the Same Mean; the Thin S-Shaped Curve is that of the Normal or Gaussian, the Thick Curve is the Exponential.

4.2.3. Exponential Distribution

If the number of events in a given interval has the Poisson distribution, then the time t between events has the *exponential* distribution $F[t|\lambda] = 1 - e^{-\lambda t}$ for $t > 0$ that is depicted in Figure 4.1.

Now, imagine a system on a spacecraft, for example, where various critical components have been duplicated, so that k consecutive failures are necessary before the system as a whole fails. If each component has a lifetime that follows the exponential distribution $F[t|\lambda]$, then the lifetime of the system as a whole obeys the chi-square distribution with k degrees of freedom.

4.3. Normal Distribution

To test the models we derived in Chapter 2, we need to make some kind of assumption about the distribution of the random component. In many, but not all cases, the frequency distribution takes the form depicted in Figure 4.2. This distribution is

a. symmetrical about the mode;
b. its mean, median, and mode are the same;
c. most of its values, approximately 68%, lie within one standard deviation of the median, 95% lie within two standard deviations;
d. yet arbitrarily large values are possible, but with vanishingly small probability.

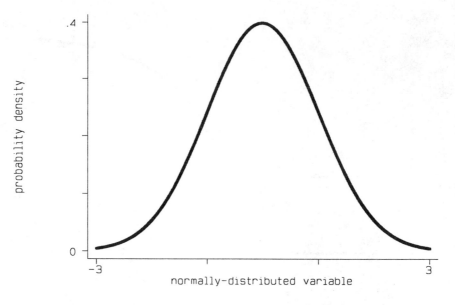

FIGURE 4.2. Probability Distribution of a Normally Distributed Variable.

Such a distribution, termed a "normal" or "Gaussian" distribution, arises whenever each value is actually the sum of a large number of individual values, each of which only makes a small contribution to the total. Of course, this is just what we would expect of "errors" that cannot be attributed to any single specific cause.[3]

For example, if we make repeated measurements (a classmate's height, the length of a flower petal), we will seldom get the same value twice (if we measure to sufficient decimal places) but rather a series of measurements that are normally distributed about a single central value.

An interesting and useful property of normally distributed observations is that their sum also has a normal distribution. In particular, if ϵ_1 is normal with mean μ_1 and variance σ_1^2, $N(\mu_1, \sigma_1^2)$, and ϵ_2 is normal with mean μ_2 and variance σ_2^2, $N(\mu_2, \sigma_2^2)$, then $\epsilon_1 + \epsilon_2$ is normal with mean $\mu_1 + \mu_2$ and variance $\sigma_1^2 + \sigma_2^2$.

4.3.1.   Tests for Location

A parametric test for location based on tables of the normal distribution is seldom used in practice, because the normal $N(\mu, \sigma^2)$ involves not one but two parameters and the population variance σ^2 is almost never known with certainty.

---

[3] Recall that we defined the arithmetic mean to be the sum of the observations divided by their number. This means that as the sample grows very large, the contribution of any individual observation $\frac{x_i}{n}$ becomes exceedingly small. Not surprisingly, the limiting distribution of the mean is the normal distribution.

To circumvent this difficulty, W. E. Gosset, a statistician with Guinness Brewery, proposed the following statistics: The test for $\mu = \mu_0$ for unknown variance is based on the distribution of

$$t = \frac{(\bar{x} - \mu_0)\sqrt{N}}{s} = \frac{(\bar{x} - \mu_0)\sqrt{N}}{\sqrt{\sum_{i=1}^{n}(x_i - \bar{x})^2/(n-1)}}$$

The test for $\mu_x = \mu_y$ for unknown but equal variances is based on the distribution of

$$t = \frac{(\bar{x} - \bar{y})}{\sqrt{\frac{(n_x-1)s_x^2 + (n_y-1)s_y^2}{N_x + N_y - 2}\left(\frac{1}{N_x} + \frac{1}{N_y}\right)}}.$$

The first of these statistics for the one-sample case has Student's t distribution with $n - 1$ degrees of freedom; the second, used for the two-sample comparison, has Student's t distribution with $n_x - n_y - 2$ degrees of freedom.[4] Today, of course, we don't need to know these formulas or learn how to look up the p-values in tables; a computer will do the work for us with nearly all commercially available statistics packages. For samples of eight or more observations, the t-test generally yields equivalent results to the permutation tests introduced in Chapter 3. For smaller samples, the permutation test is recommended, unless you can be absolutely sure the data is drawn from a normal distribution (see Section 6.3.4).

4.3.2. Tests for Scale

If an observation x has a normal distribution with mean 0 and variance σ^2, then x^2/σ^2 has a chi-square distribution with 1 degree of freedom. Suppose we wish to test whether the actual variance is greater than or equal to σ_0^2 against the alternative that it is less. If we compare the observed value of x^2/σ_0^2 against a table of the chi-square distribution, and we find that a value this large or larger would occur less than α of the time by chance, we will reject the hypothesis; otherwise we will accept it.

Most of the time, if we don't know what the variance is we won't know what the mean is either. We can circumvent this problem by taking a sample of n observations from the distribution and letting

$$S^2 = \frac{s^2}{\sigma_0^2} = \frac{\sum_{i=1}^{N}(x_i - \bar{x})^2/(n-1)}{\sigma_0^2}.$$

If the observations are from a normal distribution, our test statistic, which as you can see is based on the sample variance, will have the chi-square distribution with $n - 1$ degrees of freedom. We can check its observed value against tables of this

---

[4]Why Student? Why not Gossett's test? Guinness did not want other breweries to guess it was using Gossett's statistical methods to improve beer quality, so Gossett wrote under a pseudonym as Student.

distribution to see if it is extreme. A major caveat is that the observations must be normal and, apart from observational errors, most real-world data is not (see, for example, Micceri, 1989).

Even small deviations from normality will render this approximation invalid. My advice is to use one of the approaches described in Section 3.4 whenever you need to test for scale. Use the chi-square distribution only when you have samples of 30 or more.

For comparing the variances of two normal populations, the test statistic of choice is

$$F = \frac{\sum_{i=1}^{N}(x_i - \overline{x})^2/(n_x - 1)}{\sum_{i=1}^{N}(y_i - \overline{x})^2/(n_y - 1)}$$

A similar caveat applies: If you can't be sure the underlying distributions are normal, don't use this test statistic but adopt one of the procedures described in Chapter 8. Again, the exception is with large samples each of 30 or more in number.

4.4. Distributions

An observation X is said to have the distribution F if for all similar observations x, $F[x]$ is the probability that the observation takes on values less than or equal to x, that is, $F[x] = \Pr\{X \leq x\}$. F is monotone nondecreasing in x, that is, it never decreases when x increases, and $0 \leq F[x] \leq 1$.

The distribution function of the binomial $B(p, n)$ discussed in Section 4.2.1 is

$$F[k] = \sum_{j=0}^{k} \binom{n}{j} p^j (1 - p)^{n-j} \text{ for } k = 0, 1, 2, \cdots n.$$

The distribution function of the Poisson $P(\lambda)$ is

$$F[k] = \sum_{j=0}^{k} \lambda^j e^{-\lambda}/j! \text{ for } k = 0, 1, 2, \cdots.$$

Almost all errors have distributions that are made up of combinations or transformations of one or more of the normal, the binomial, and the Poisson. In most cases, it is difficult to characterize such mixed distributions by means of a formula. A simple example would be when a population is made up of two subpopulations, each of which has the normal distribution. Such a population would have a distribution that looks a great deal like Figure 4.3.

4.5. Summary and Further Readings

In this chapter, you had a brief introduction to parametric distributions including the binomial, Poisson, Gaussian, and exponential and to several distributions derived

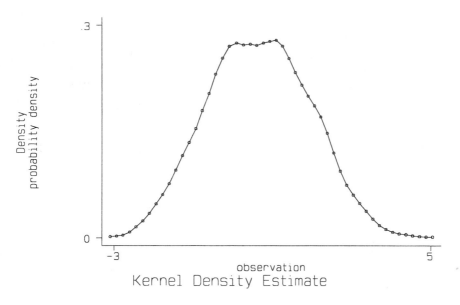

FIGURE 4.3. A Sample from a Mixture of Two Normal Distributions.

Which Distribution?

Counting the number of successes in N independent trials? Binomial.

Counting the number of rare events that occur in a given time interval or a given region? Poisson.

Recording the length of the interval that elapses between rare events? Exponential.

What you observe is the sum of a large number of factors, each of which makes only a small contribution to the total? Normal.

---

from them including χ^2, Student's t, and F. Relationships among these and other common univariate distributions are described in Leomis [1986]. You learned here how to obtain tests of hypothesis based on these distributions. For a further discussion of the underlying theory, see Lehmann [1986]. The CRC Handbook of Tables for Probability and Statistics can help you derive the appropriate cut-off values.

4.6. Exercises

1. Seventy percent of the registered voters in Orange County are Republican. What is the probability that the next two people in Orange County you ask about their political preferences will say they are Republican? The next two out of three

people? What is the probability that at most one of the next three people will admit to being a Democrat?

2. Two-thousand bottles of aspirin have just rolled off the assembly line; 15 have ill-fitting caps, 50 have holes in their protective covers, and 5 are defective for both reasons. What is the probability that a bottle selected at random will be free from defects? What is the probability that a group of 10 bottles selected at random will all be free of defects?

3. For each of the following, indicate if the observation is binomial, Poisson, exponential, normal or almost normal, comes from some other distribution, or is predetermined (that is, not random at all).

   a. Number of books in your local public libarary.
   b. Guesses as to the number of books in your local public library
   c. Heights of Swedish adults.
   d. Weights of Norwegian women.
   e. Alligators in an acre of Florida swampland.
   f. Vodka drinkers in a sample of 800 Russians.
   g. Messages on your answering machine.
   h. You shoot at a target; the distance from the point of impact to the target's center.

4. How many students in your class have the same birthday as the instructor? In what proportion of classes, all of size 20, would you expect to find a student who has the same birthday as the instructor? Two students who have the same birthday?

5. What is the largest number of cells you can drop into a Petri dish divided into 100 squares and be sure the probability a square contains two or more cells is less than 1%? What percentage of the squares would you expect would contain exactly 1 cell?

6. Bus or subway? The subway is faster but there are more buses on the route. One day, you and a companion count the number of arrivals at the various terminals: 14 buses arrive in the same period as only 4 subway trains pull in. Use the binomial distribution to determine whether the bus is a significantly better choice than the subway.

7. Reanalyze the data of Exercises 3.12 and 3.13, assuming the data in these exercises came from a normal distribution.

8. Show that the variance of a binomial with n trials and probability p of success of each trial is $np(1 - p)$.

9. A poll taken in January 1998 of 1,000 people in the United States revealed that 64% felt President Clinton should remain in office. Suppose that 51% is the true value, what would be the variance of this estimate?

CHAPTER 5

Estimation

In this chapter, you'll expand on the knowledge of estimation you gained in Chapter 1 to make point and interval estimates of the effects you detected in Chapter 3.

5.1. Point Estimation

"Just give me the bottom line," demands your supervisor, "one number I can show my boss." The recommended point, or single-value estimate, in most cases, is the *plug-in* estimate: For the population mean, use the sample mean; for the population median, use the sample median; and to estimate a proportion in the population, use the proportion in the sample. These plug-in estimates have the virtue that they are *consistent*, that is, as the sample grows larger, the plug-in estimate grows closer to the true value. In many instances, particularly for measures of location, plug-in estimates are unbiased, that is, they are closer on the average to the true value than to any other value.

The exception that proves the rule is the population variance, which you'll consider in Exercise 6. The recommended estimate, recommended because it is unbiased, is $\frac{1}{n-1}\sum_{i=1}^{n}(x_i - \bar{x})^2$, while the plug-in estimate is $\frac{1}{n}\sum_{i=1}^{n}(x_i - \bar{x})^2$.

It is customary when reporting results to provide both the estimate and an estimate of its standard error.[1] The bootstrap will help us here. We pretend the sample is the population and take a series of bootstrap samples from it. We make an estimate for each of these samples; label this estimate $\widehat{\theta_b}$. After we have taken B

---

[1]Customary, but not always sensible; see the next section.

bootstrap samples, we compute the standard error of our sample of B estimates

$$\sqrt{\frac{1}{B-1}\sum_{b=1}^{B}(\widehat{\theta}_b - \overline{\widehat{\theta}}_.)^2}$$

where $\overline{\widehat{\theta}}_. = \frac{1}{B}\sum_{b=1}^{B}\widehat{\theta}_b$ is the mean of the bootstrap estimates.

Suppose we want to make an estimate of the median height of Southern California sixth-graders based on the heights of my sixth-grade class. Our estimate is simply the median of the class, 153.5 cm. To estimate the standard error, or *precision* of this estimate, we take a series of bootstrap samples from the heights recorded in Table 1.1. After reordering each sample, we have

Bootstrap Sample 1:   137.0   138.5   138.5   141.0   142.0   145.0   145.0
147.0   148.5   148.5   150.0   150.0   154.0   155.0   156.5   157.0
158.0   158.5   159.0   160.5   161.0   161.0;   Median = 150.0

Bootstrap Sample 2:   138.5   140.0   140.0   141.0   142.0   143.5   145.0
148.5   150.0   150.0   153.0   154.0   156.5   157.0   158.0   158.5
159.0   160.5   161.0   162.0   162.0   167.5;   Median = 151.5

Bootstrap Sample 3:   140.0   141.0   142.0   143.5   145.0   147.0   148.5
150.0   153.0   155.0   156.5   156.5   157.0   158.0   158.5   159.0
159.0   159.0   160.5   161.0   162.0   167.5;   Median = 155.75

If we were to stop at this point (normally, we'd take at least 50 bootstrap samples), our estimate of the standard error of our estimate of the median would be

$$\sqrt{\frac{(150 - 152.8)^2 + (151.5 - 152.8)^2 + (155.75 - 152.8)^2}{3-1}}.$$

5.2.   Interval Estimation

The standard error is only an approximation to the actual dispersion of an estimate. If the observations are normally distributed (see Section 4.3) or if the sample is large, then the interval from one standard error below the sample mean to one standard error above it will cover the true mean of the population about two-thirds of the time. But many estimates (for example, that of a variance or a correlation coefficient) do not have a symmetric distribution so that such an interpretation of the standard error would be misleading. In the balance of this section, you learn to derive a more accurate interval estimate, one hopefully more likely to cover the true value than a false one.

5.2.1.   An Untrustworthy Friend

We begin our investigation of improved interval estimates with the simplest possible example, a binomial variable.

My friend Tom settles all his arguments by flipping a coin. So far today, he's flipped five times, called heads five times, and got his own way five times. My natural distrust of my fellow humans leads me to suspect that Tom's not using a fair coin. Recall from Section 2.2.2.2, that if p is the probability of throwing a head,

$$\Pr\{5 \text{ heads in a row}\,|\,p\} = p^5(1-p)^0.$$

If $p = 0.5$, as it would for a fair coin, this probability is 0.032. Awfully small. If $p = 0.65$, this probability is 11%. Could be. In fact, for any value of $p > .55$, this probability is at least 5%. But not for $p = .5$. Do you suppose Tom's coin is weighted so that it comes up heads at least 55% of the time? No. He wouldn't do that. He's my friend.

In order to be able to find a *lower confidence bound* on p for any desired significance level, I built the chart shown in Figure 5.1. To determine a lower bound, first find the significance level α on the y-axis; then read across to the curve and down to the x-axis. For example, if $\alpha = .05$, then $p > .55$.

This chart is applicable not only to suspect coins but to any example involving variables that have exactly two possible outcomes.

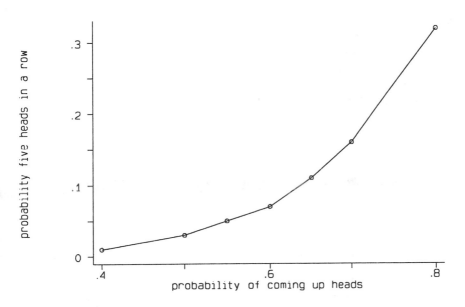

FIGURE 5.1. Deriving Lower Confidence Bounds for the Binomial.

5.2.2.   Confidence Bounds for a Sample Median

In this example, we have a set of observations $-1, 2, 3, 1.1, 5$ taken from a much larger population and we want to find a *confidence interval* that will cover the true value of the unknown median, θ, $1 - \alpha$ of the time. As an alternative to the bootstrap described in Section 1.5.4, the method of randomization can help us find a good interval estimate.

Recall from Section 3.6.2 that in the first step of our permutation test for the location parameter of a single sample, $H : \theta = \theta_o$ versus $K : \theta > \theta_o$ we subtract θ_o from each observation. We might test a whole series of hypotheses involving different values for θ_o until we find a θ_1 such that as long as $\theta_o \geq \theta_1$, we accept the hypothesis at the α level, but if $\theta_o < \theta_1$ we reject it. Then $1 - \alpha$ confidence interval for θ is given by the one-sided interval $\{\theta \geq \theta_1\}$.

Let's begin by testing the hypothesis $\theta = 0$. Our first step is to compute the sum of the negative[2] deviations about 0, which is -1. Among the $2 \times 2 \times 2 \times 2 \times 2 = 2^5$ possible reassignments of plus and minus signs are

$$+1, \quad -2, \quad +3, \quad +1.1, \quad +5$$
$$+1, \quad +2, \quad +3, \quad +1.1, \quad +5$$

and

$$-1, \quad -2, \quad +3, \quad +1.1, \quad +5$$

Our next step is to compute the sum of negative deviations for each rearrangement. For the three arrangements shown here, this sum would be -2, 0, and -3, respectively. Only 2 of the 32 possible rearrangements, or $1/16$, result in samples that are as extreme as that of the original observations. We accept the hypothesis that $\theta \leq 0$ at the $1/16$ level and thus, at any smaller α-level including the $1/32$. Similarly, we accept the hypothesis that $\theta \leq -0.5$ at the $1/32$ level, and even that $\theta \leq -1 + \epsilon$ where ϵ is an arbitrarily small but still positive number. But we would reject the hypothesis that $\theta \leq -1 - \epsilon$, because after subtracting $-1 - \epsilon$ the transformed observations $\epsilon, 2, 3, 1.1, 5$ are all larger than zero.

Our one-sided confidence interval is $\{\theta > -1\}$, and we have confidence that $31/32$ of the time this method yields an interval that includes the true value of the location parameter θ. [3]

Our one-sided test of a hypothesis gives rise to a one-sided confidence interval. But knowing that θ is larger than -1 may not be enough. We may want to pin θ down to a more precise two-sided interval, say that θ lies between -1 and $+1$.

To accomplish this, we need to begin with a two-sided test. Our hypothesis for this test is that $\theta = \theta_o$ against the two-sided alternative that θ is either smaller or

---

[2] I chose negative deviations rather than positive because there are fewer of them and thus fewer calculations, at least initially.

[3] It is the method that inspires this degree of confidence; a second sample might yield a quite different interval.

larger than θ_o. We use the same test statistic, the sum of the negative observations, that we used in the previous one-sided test. Again, we look at the distribution of our test statistic over all possible assignments of the plus and minus signs to the observations. But this time we reject the hypothesis if the value of the test statistic for the original observations is either one of the largest or one of the smallest of the possible values.

In our example, we don't have enough observations to find a two-sided confidence interval at the 31/32 level, so we'll try to find one at the 15/16. The lower boundary of the new confidence interval is still -1, but what is the new upper boundary? If we subtract 5 from every observation, we would have the values -6, -3, -2, -3.9, and -0; their sum is -15.9. Only the current assignment of signs to the transformed values, that is, only one of the 32 possible assignments yields this large a sum for the negative values. The symmetry of the two-sided permutation test requires that we set aside another 1/32 of the arrangements at the high end. Thus we would reject the hypothesis that $\theta = 5$ at the $1/32 + 1/32$, or 1/16 level. Consequently, the chances are 15/16 that the interval $\{-1, 5\}$ covers the unknown parameter value.

These results are readily extended to a confidence interval for a *vector* of parameters θ that underlies a one-sample, two-sample, or k-sample experimental design with single- or vector-valued variables. For example, we might want a simultaneous estimate of the medians of both arm span and height, so that $\theta = (\theta_1, \theta_2)$. The $1 - \alpha$ confidence interval consists of all values of the parameter vector θ for which we would accept the hypothesis at level α. Remember, one-sided tests produce one-sided intervals, and two-sided tests produce two-sided confidence intervals.

5.2.3. A Large-Sample Approximation

We can derive an approximate confidence interval for a median quickly by using the ranks of the signed observations. Let's take another look at the process temperature data in Table 3.2. Entering this data into StatXact-3 as CaseData, we pull down successive menus corresponding to **Statistics, Paired-Samples,** and the **Hodges-Lehmann Confidence Interval for Shift** described in Hodges and Lehmann [1963].

Lower Confidence Bound for a Median

I. Permutation Approach

Guess at a lower bound L.
Test the hypothesis that the Median is L.
   If you accept this hypothesis, decrease the bound and test again.
   If you reject this hypothesis, increase the bound and test again.

HODGES-LEHMANN ESTIMATES OF MEDIAN DIFFERENCE
Summary of Exact Distribution of WILCOXON SIGNED RANK

Min	Max	Mean	Std-dev	Observed	Standardized
0.0000	210.0	105.0	26.78	210.0	3.921

Point Estimate of Mean Process Temperature = 455.5

95.00% Confidence Interval for Mean Process Temperature:
    Asymptotic : (446.0,465.0)

5.2.4. Confidence Intervals and Rejection Regions

There is a direct connection between confidence intervals and the rejection regions of our tests. Suppose $A(\theta')$ is a $1 - \alpha$ level acceptance region for testing the hypothesis $\theta = \theta'$, that is, we accept the hypothesis if our test statistic $x \in A(\theta')$ and reject it otherwise. Suppose $S(X)$ is a $1 - \alpha$ level confidence interval for θ based on the set of observations $X = \{x_1, x_2, \cdots x_n\}$. Then, $S(X)$ consists of all the parameter values $\theta*$ for which X belongs to the acceptance region $A(\theta*)$, while $A(\theta)$ consists of all the values of the statistic x for which θ belongs to the confidence interval $S(x)$.

$$\Pr\{S(X) \text{ includes } \theta_o \text{ when } \theta = \theta_o\} = \Pr\{X \in A(\theta_o) \text{ when } \theta = \theta_o\} \geq 1 - \alpha.$$

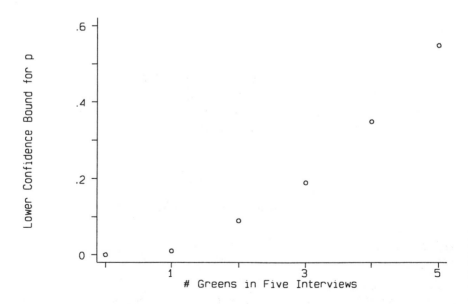

FIGURE 5.2. Lower Confidence Bound as a Function of the Number of Positive Responses in 5 Binomial Trials.

Important Terms

Acceptance region, $A(\theta_o)$: set of values of the variable X for which we accept the hypothesis $H : \theta = \theta_o$. Its complement is called the rejection region.

Confidence region, $S(X)$: Also referred to as a *confidence interval* (for a single parameter) or a *confidence ellipse* (for multiple parameters); set of values of the parameter θ for which, given the set of observations $X = \{x_1, x_2, \cdots, x_n\}$, we would accept the corresponding hypothesis.

To illustrate the duality between confidence bounds and rejection regions, we've built the curve shown in Figure 5.2. (See Exercise 3 for the details of the construction.) Now, let's take a second look at the problem of the unfair coin that we considered at the beginning of this chapter. Instead of tossing a coin, let's consider a survey in which we ask 5 individuals selected at random whether they plan to vote for the Green candidate. We use the number of positive responses in the 5 interviews to obtain a lower confidence bound for the probability p that an individual selected from the population at large will vote for the Green candidate. Turning to Figure 5.2 and entering the chart from below along the horizontal axis, we see that if 2 of the 5 favor the Green, a 95% lower confidence bound for p is .09. Alternatively, if we wish to test the hypothesis that $p \leq .20$, we enter the chart from the left side along the vertical axis to learn we should reject at the 5% level if three or more of the selected individuals favor Green.

5.3. Regression

Let's take a second look at the conference data we gathered in Section 2.2.1. with the aim of deriving point and interval estimates for the parameters of the models we developed there. Recall that as the new owner of Fawlty Towers, you are trying to cut costs by preparing only as many meals as will be required. The experience of your predecessor, summarized in Table 5.1, was that many of the registrants would prove to be no-shows, and the number of no-shows, and more specifically, of nondiners would increase dramatically in bad weather. Once the conference officials supply the number of registrants and you've had a chance to look at the weather forecast a regression equation of the form

$$\text{Meals} = a + b * \text{Registrants}$$

or

$$\text{Meals} = a + b * \text{Registrants} + c * \text{Weather}$$

will allow you to predict the expected number of meals.

TABLE 5.1. Conference Registrants, Meals Served, and Weather at Fawlty Towers

Registrants	Max Servings	Weather	Registrants	Max Servings	Weather
289	235	3	339	315	2
391	355	3	479	399	2
482	475	1	500	441	2
358	275	4	160	158	3
365	345	1	319	305	2
561	522	1	331	225	5

I'll assume you have a loss function that depends on the difference between the predicted and the actual number of meals you want to minimize.[4] A common form for such a function is $(M' - M)^2$ where M' is your prediction and M is the actual number of meals. If you've taken calculus, then you know the minimum of this function for the data at hand is obtained for the first equation when

$$\Sigma(a + bR_i - M_i)b = 0 \text{ and } \Sigma(a + bR_i - M_i) = 0$$

that is, when

$$b = \frac{\text{Covariance}(RM)}{\text{Variance}(M)} = \frac{\Sigma(R_i - \overline{R})(M_i - \overline{M})}{\Sigma(M_i - \overline{M})^2} \text{ and } a = \overline{M} - b\overline{R}.$$

Those with a knowledge of matrix algebra can also derive the optimal values of the parameters a, b, and c for the second equation.[5] I prefer to let a program like SPSS or Stata do the work for me. The *least-square*-loss solution to this regression equation is Meals $= 104 + .784 * \text{Registrants} - 27 * \text{Weather}$.

Thus, if there are 300 registrants and the weather forecast calls for clear skies (weather $= 1$), you would predict $104 + 234 - 27 = 211$ meals should be served and plan accordingly.

The preceding discussion assumes we use the regression equation for *interpolation* within the range of known values. We are on shaky ground if we try to *extrapolate*, to make predictions for conditions not previously investigated, like Weather $= 1$ and Registrants $= 100$. Meals for 100 registrants $= 104 + 78 - 27 = 155$? I don't think so.[6]

---

[4] See Section 5.4 for more on the economics of decision making.

[5] Let β denote the vector of coefficients (a, b, c), M the vector of servings $(M_1, \ldots M_n)$, and C the $n \times 3$ matrix whose ith row is $(1, R_i, W_i)$, then the least-squares estimate of β is $(C^TC)^{-1}C^TM$.

[6] Extrapolation, inevitable in survival and reliability analysis, will be considered in Chapter 9.

5.3.1. Prediction Error[7]

For most hotel owners (Basil Fawlty excepted) a single figure would not be enough; a prediction interval is needed. Averages are often exceeded and the sight of six or seven conference participants fighting over a single remaining pork cutlet can be a bit unsettling. In this section, we bootstrap not once, but twice to obtain the desired result.

Consider the more general regression problem where given a vector x_i of observations, we try to predict y_i. Suppose we've already collected a set of independent identically distributed observations $w = \{w_i = (x_i, y_i), i = 1, \ldots n\}$ taken from the multidimensional distribution function F. The data in Table 5.1, with x the number of registrants, and y the number of meals provides one example. Given a new value for the number of registrants x_{n+1}, we use our regression equation based on w to derive an estimate $\eta[x_{n+1}, w]$ of the number of meals y_{n+1} we can expect to serve. Any difference between the estimate and the actual number of meals served could result in a loss. The loss could be something as simple as the cost $\$A$ of preparing an unnecessary meal or it could be far more complicated. For example,

$$
L[y_{n+1}, \eta[x_{n+1}, w]]
$$
$$
= \begin{cases} (\eta[x_{n+1}, w] - y_{n+1})\$A & \text{if } y_{n+1} < \eta[x_{n+1}, w] \\ (y_{n+1} - \eta[x_{n+1}, w])\$B & \text{if } y_{n+1} > \eta[x_{n+1}, w] \\ (y_{n+1} - \eta[x_{n+1}, w])\$B + \$C & \text{if } y_{n+1} > \eta[x_{n+1}, w] + 10 \end{cases}
$$

where $\$B$ the cost of a replacement meal obtained from a nearby hotel and $\$C =$ the costs of the law suits brought by thoroughly irritated porcupine fanciers.

The prediction error $L[y_i, \eta[x_i, w]]$ should be averaged over all possible outcomes (x_{n+1}, y_{n+1}) we might draw from F in the future. Call the result $\text{eff}[w, F]$. But we don't know what F is! We do have an estimate, F', the empirical distribution function, based on the data w we've collected so far. The apparent error rate is the average of the losses over all the pairs of values (x_i, y_i) we have observed:

$$
\text{eff}[w, F'] = \frac{1}{n} \sum_{i=1}^{n} L[y_i, \eta[x_i, w]].
$$

If we choose as loss function, $L = |y_i - \eta[x_i, w]|$, then the apparent error rate is the sum of the absolute values of the residuals.

To estimate the true error, we treat the observations as if they were the population and a series of bootstrap samples as if they were the observations. Each bootstrap sample consists of pairs $\{w_i*\}$ drawn without replacement from w. Our plug-in

---

[7]This section and the next will be challenging for the algebraically disadvantaged. Two alternatives suggest themselves: 1) read through the sections quickly just to get a feel for the type of problems they can help solve, then come back if and when you need the help; 2) work through the algebra step by step substituting real data.

estimate of error based on a bootstrap sample is

$$\text{eff}[w*, F'] = \frac{1}{n} \sum_{i=1}^{n} L[y_i, \eta[x_i, w*]].$$

To obtain this estimate, we must compute the regression coefficients a second time using $w*$ in place of w in our calculations, although we continue to average the errors over the original observations. Because results based on a single bootstrap sample can be quite misleading, we have to repeat the process of sampling, regression, and error determination several hundred times, and then take the average of our estimates. I'm not complaining because, like you, I have a computer to do the work. Our final estimate of prediction error is the average of the plug-in estimates over the B bootstrap samples,

$$\text{eff}[w*, \widehat{F}] = \frac{1}{B} \sum_{b=1}^{B} \left(\frac{1}{n} \sum_{i=1}^{n} [y_i, \eta[x_i, w^{*b}]] \right).$$

We use the distribution of errors from these same calculations to obtain an interval estimate of the prediction error.

5.3.2. Correcting for Bias

When we estimate a population mean using the sample mean, the result is *unbiased*, that is, the mean of the means of all possible samples taken from a population is the *population mean*. In the majority of cases, including the preceding example, estimates of error based on bootstrap samples are biased and tend to underestimate the actual error. If we can estimate this bias, bootstrapping a second time, we can improve on our original estimate.

This bias, which Efron and Tibshirani [1993] call the *optimism*, is the difference $\text{eff}[w, F] - \text{eff}[w, \widehat{F}]$, which we estimate as before by using the observations in

---

Estimating Prediction Error I

Determine the loss function L.
Repeat several hundred times.
    Draw a bootstrap sample $w*$ from the observation pairs

$$w = \{(x_i, y_i)\}$$

    Compute the regression coefficients based on this sample. Using these coefficients, compute the losses $L[y_i, \eta[x_i, w*]]$ for each pair (x_i, y_i) in the original sample.
    Compute the mean of these losses.
Compute the mean of these means averaged over all the bootstrap samples.

Estimating the Prediction Error II

Determine the loss function L.
Solve for the regression coefficients using the original observations
Compute apparent error rate:

$$\text{eff}[w, F] = \frac{1}{n} \sum_{i=1}^{n} L[y_i, \eta[x_i, w]]$$

Repeat the following 160 times:

1. Choose a bootstrap sample $w*$ with replacement.
2. Solve for regression coefficients using the bootstrap sample.
3. Use these coefficients to determine the apparent error for original observations

$$\text{eff}[w*, \widehat{F}*]$$

and the optimism

$$\text{eff}[w*, \widehat{F}] - \text{eff}[w*, \widehat{F}*]$$

Compute the average optimism.
Compute the corrected error rate.

---

place of the population and the bootstrap samples in place of the observations,

$$\text{eff}[w*, \widehat{F}] - \text{eff}[w*, \widehat{F}*] = \frac{1}{Bn} \sum_{b=1}^{B} \sum_{i=1}^{n} (L[y_i, \eta[x_i, w^{*b}]] - L[y_i^{*b}, \eta[x_i^{*b}, w^{*b}]]).$$

$L[y_i, \eta[x_i, w^{*b}]]$ uses the observations $\{(x_i, y_i)\}$ from the original sample and the coefficients derived from the bootstrap; $L[y_i^{*b}, \eta[x_i^{*b}, w^{*b}]]$ uses the bootstrap observations $\{(x_i^{*b}, y_i^{*b})\}$ and the coefficients derived from the bootstrap. Our corrected estimate of error is the apparent error rate plus the optimism or

$$\text{eff}[w, \widehat{F}] + \text{eff}[w*, \widehat{F}] - \text{eff}[w*, \widehat{F}*].$$

5.4. Which Model?

Suppose we are confronted with several explanations for the same set of data.

$$\text{Model I} \qquad y = ax + \epsilon$$
$$\text{Model II} \qquad y = ax^2 + \epsilon$$

or even

$$\text{Model III} \qquad y = a \log(x_1) + \epsilon$$

or

$$\text{Model IV} \qquad y = a_1 x_1 + a_2 x_2 + \epsilon.$$

Is the best model the one that produces the smallest residual errors, so-called *goodness-of-fit*? Not necessarily; our object is not merely to fit the data at hand but to successfully predict future outcomes.

Perhaps we could divide the data in half and use one half to develop our models and the other to validate and compare them. But which half? And won't we be throwing away valuable information?

Our answer is to use all the data for both purposes, one purpose at a time. With the delete-50% cross-validation method first proposed by Geisser [1975], we divide our sample of size $2n$ into two equal parts at random.[8] We use the first part of the data, call this set S, to estimate the matrix of model parameters A_S and the second part S^c to check the accuracy of the resultant model, that is, in symbols, to determine the prediction error,

$$\sum_i \epsilon S^c L(y_i - A_S X_i)$$

For example, with the data from Fawlty Towers, we might use the six pairs of observations (289,235), (391,355), (358,275), (339,315), (160,159), (319,305) to estimate the regression coefficients for each competing model, and the errors when we try to fit the remaining six pairs to determine the prediction errors.

We repeat this process 20 or 30 times—selecting six points, estimating the regression coefficients, computing the prediction errors. Our model of choice is the one that minimizes the average:

$$\frac{1}{B} \sum_{j=1}^{B} \sum_i \epsilon S_j^c L(y_i - A_{S_j} X_i)$$

where $B = 20$ or 30 or whatever number of resamples you chose.

5.5. Bivariate Correlation

When we test for an ordered dose response, our underlying model is that Y, the observed effect, depends in a linear fashion on the logarithm of the dose X; that is, $Y = a + b. \log X$, or, if F and G are the corresponding frequency distributions, that $F(y|X) = G(y - b. \log[X])$. The test for $b = b_0$ based on Pitman correlation is described in Section 3.5. A confidence interval for b is readily derived using permutation methods (see Exercise 4) providing we can assume the errors are symmetrically distributed.

In the case of bivariate correlation, although we may perform an initial test to determine whether $Y = bX$ with $b \neq 0$, our primary concern is with the proportion

---

[8] We can take advantage of the same computer program we used in Chapter 3 to obtain a random rearrangement of two samples.

FIGURE 5.3. Results of 400 Bootstrap Simulations Using Table 5.2

of the observation Y that is explained by X. One measure of this proportion and the degree of association between X and Y is the correlation ρ defined as

$$\rho = \frac{\text{Cov}(XY)}{\sigma_x \sigma_y}$$

or

$$\rho = b \frac{\sigma_x}{\sigma_y}$$

where the covariance, when two variables, x and y, are observed simultaneously, is defined as

$$\text{Cov}(XY) = \sum_{i=1}^{n} (\overline{x} - x_i)(\overline{y} - y_i)/(n-1)$$

If X and Y are independent and uncorrelated, $\rho = 0$. If X and Y are totally dependent, for example, if $X = 2Y|\rho| = 1$. In most cases of dependence $0 < |\rho| < 1$.

If we knew the standard deviations of X and Y, we could use them to help obtain a confidence interval for ρ by the method of permutations. But we don't know σ_x and σ_y; rather we know their estimates, just as we only have an estimate of b.

Fortunately, we can use the bootstrap to obtain a confidence interval for ρ in a single step, just as we did in Section 1.5.4. Does a knowledge of algebra ensure a better grade in statistics? Table 5.2 displays test score data from a study by Mardia, Kent, and Bibby [1979]. A box plot of the bootstrap results using a Stata routine (see Appendix 3) is depicted in Figure 5.3. The median correlation is .68, with a 90% confidence interval of (.50, .87). This interval does not include 0, so we reject the null hypothesis and conclude that a knowledge of algebra does indeed ensure a better grade in statistics (see Section 5.2.2). See, also, Young [1988] .

TABLE 5.2. Score Data from Mardia, Kent, and Bibby [1979]

Algebra	67	80	71	63	65	72	65	68	58	60	60	59
Statistics	81	81	81	68	63	73	68	56	70	45	54	44

Confidence Interval for the Correlation Coefficient

Derive the bootstrap distribution.

1. Choose a bootstrap sample without replacement.
2. Compute the correlation coefficient.
3. Repeat several hundred times.

Select a confidence interval from this distribution.

5.6. Smoothing the Bootstrap

This balance of this chapter will be challenging for the algebraically disadvantaged. Two alternatives suggest themselves: 1) read through the next few sections quickly to get a feel for the type of problems they can solve and come back to them if and when you need them; or 2) work through the algebra step by step, substituting real data for symbols.

An inherent drawback of the bootstrap, particularly with small samples, lies in the discontinuous nature of the empirical distribution function. Our samples presumably come from a continuous or near-continuous[9] distribution; Figure 5.4 illustrates the distinction. The jagged curve is the empirical distribution function; the smooth curve that passes through it was obtained by replacing the original observations with their Studentized equivalents[10], $z_i = (x_i - \bar{x})/s$, then plotting $N(z_i)$ where N is the distribution function of a normally distributed random variable, $N(\bar{x}, s^2)$.

To obtain an improved bootstrap interval for the median height of a typical sixth-grader based on the data in Table 1.1, we modify our bootstrap procedure as follows:

1. Sample $z_1, z_2, \ldots z_n$ with replacement from the original observations $x_1, x_2, \ldots x_n$.
2. Compute their mean $\bar{z}$ and the plug-in estimate of the population variance $\hat{\sigma}^2 = \sum_{i=1}^{n}(z_i - \bar{z})^2/n$.
3. Set

$$x_i* = \bar{z} + \frac{(z_i - \bar{z} + h\epsilon_i)}{\sqrt{1 + h^2/\hat{\sigma}^2}}$$

where the ϵ_i are drawn from an $N(0, 1)$ distribution, and h is a smoothing constant, chosen by trial and error,
4. Use $x_1^*, x_2^*, \ldots x_n^*$ to estimate the median.

To obtain a smoothed bootstrap estimate of the distribution of the correlation coefficient, we use a similar, albeit more complex, procedure:

---

[9] In the current example, our observations are test scores that are discrete integers.
[10] Named after Student, the pseudonym of the statistician W.S. Gosset.

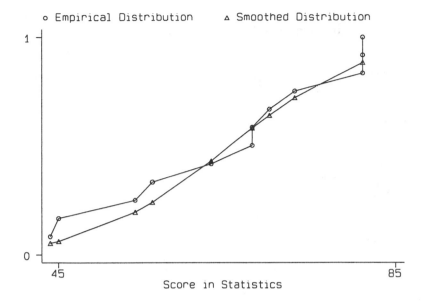

FIGURE 5.4. Cumulative Distribution Function of Statistics Scores, Derived from Table 5.1. The Jagged Curve o–o–o is the Empirical (Observed) Distribution Function.

1. We take a bootstrap sample $(x, y)^*_1, (x, y)^*_2, \ldots (x, y)^*_n$ with replacement from the original set of observations $(x, y)_1, (x, y)_2, \ldots (x, y)_n$.
2. Following Hall and Wilson [1991], we compute the mean of the bootstrap sample, $\mu'^* = (x\bar{*}, y\bar{*})$ and its covariance matrix Σ^*.
3. We draw n pairs (ϵ_i, η_i) from the bivariate normal distribution with mean μ'^* and covariance matrix Σ^* and use them to form

$$x_i* = x\bar{*} + \frac{(x_i - h\epsilon_i)}{\sqrt{1 + h^2/\sigma'_x}}, \; y_i^* = \bar{y} * + \frac{(y_i - h\eta_i)}{\sqrt{1 + h^2/\sigma'_y}}.$$

4. Finally, we use $(x*_1, y*_1), (x*_2, y*_2) \ldots (x*_n, y*_n)$ to estimate the correlation coefficient.

5.7. Bias-Corrected Bootstrap

The percentile bootstrap interval described in Section 1.5 and used here is not unbiased and may not be even for very large sample sizes. A so-called 95% confidence interval may only cover the correct parameter value 90% of the time. (Worse, it may cover an incorrect value of the parameter more often than it covers the correct value.) The BC_a interval[11] represents a substantial improvement for

---

[11] BC_a stands for bias-corrected and accelerated.

Computing a Smoothed Bootstrap Estimate of the Median
Sample Calculation with Stata

. bsample _N

1.	143.5	2. 160.5	3. 155	4.	137					
5.	156.5	6. 158	7. 161	8.	148.5					

. list height

9.	153	10.	141	11. 145	12.	140	
13.	147	14.	154	15.	138.5	16.	159
17.	155	18.	158	19.	141	20.	141
21.	145	22.	138.5				

$\bar{z} = 148.9$ and $\hat{\sigma}^2 = 64.18$.

. invnorm (uniform)

$$\epsilon_1 = 0.12$$

If we take $h = 0.5$ then

$$x_1^* = 148.9 + \frac{143.5 - 148.9 + 0.5\epsilon_1}{(1 + 0.25/64.18)^{1/2}} = 143.54$$

---

deriving one-sided confidence intervals[12], but for samples under size 20, it is still suspect. The idea behind these intervals is that if θ is the parameter we are trying to estimate and $\hat{\theta}$ is the estimate, we might be able to come up with an increasing transformation m such that $\phi = m(\theta)$ and $\hat{\phi} = m(\hat{\theta})$ yields $z = \frac{\phi - \hat{\phi}}{e(\hat{\phi}, a)}$, where the distribution of z is very nearly normal. We could use this normal distribution to obtain an unbiased confidence interval and then apply a back-transformation to obtain an almost-unbiased confidence interval. The method is not as complicated as it reads because we don't actually have to go through all these steps, we merely agree we could if we need to. SPlus code for obtaining BC_a intervals may be obtained from the statistics archive of Carnegie–Mellon University; send an email to statlib@lib.stat.cmu with the one-line message send bootstrap.funs from S. The equivalent SAS code is provided in Appendix 3.

5.8. Iterated Bootstrap

Bootstrap iteration provides another way to improve the accuracy of bootstrap confidence intervals. Let I_θ be a $1 - \alpha$ level confidence interval for θ whose actual coverage is $\pi_\theta(\alpha)$. In most cases $\pi_\theta(\alpha)$ will be larger than $1 - \alpha$. Let β_α be the value of α for which

$$\pi_\theta(\beta_\alpha) = P\{\theta \epsilon I_\theta(\beta_\alpha | \{X_i\})\} = 1 - \alpha,$$

---

[12]For deriving improved two-sided confidence intervals, see the discussion by Loh [1991; pages 972–976], following Hall [1988].

that is, an interval estimate that encompasses β_α of the bootstrap distribution has probability $1 - \alpha$ of covering the true parameter. We can't solve for β_α directly, but we can obtain a somewhat more accurate confidence interval by using $I_\theta(\hat{\beta}_\alpha|\{X*_i\})$, where $\hat{\beta}_\alpha$ is the estimate of β_α obtained by replacing θ with $\hat{\theta}$ and the sample $\{X_i\}$ with the bootstrap sample $\{X*_i\}$. Details of the procedure are given in Martin [1990, pp. 1113–4]. The iterated bootstrap, while straightforward, is computationally intensive. Suppose the original sample has 30 observations and we take 300 bootstrap samples from each of 300 bootstrap samples. That's 2,700,000 values, plus an equal number of calculations![13]

5.9.   Summary

In this chapter, you learned how to make point estimates and how to estimate their precision using the bootstrap. You learned how to make interval estimates using either the bootstrap or the permutation test and to improve the bootstrap estimates by a variety of methods. You learned to use the bootstrap to compare models in terms of both prediction error and goodness of fit.

5.10.   To Learn More

For further information on deriving confidence intervals using the randomization approach see Lehmann [1986, pp 246-263], Gabriel and Hsu [1983], Garthwaite [1996], John and Robinson [1983], Maritz [1996, p7, p25], Martin-Lof [1974], Noether [1978], Tritchler [1984], Buonaccorsi [1987]. Solomon [1986] considers the application of confidence intervals in legal settings. For a discussion of the strengths and weaknesses of pivotal quantities, see Berger and Wolpert [1984].

Vecchia and Iyer [1989] and van der Voet [1994] use permutation methods to compare the predictive accuracy of alternate models. Forsythe, Engleman and Jennrich [1973] develop a stopping rule for variable selection in multivariate regression. Garbiel and Hall [1983], Raz [1990] and Cade [1996] apply permutation methods to nonparametric regression, while Hardle and Bowman [1988] use the bootstrap.

For examples of the wide applicability of the bootstrap method, see Chapter 7 of Efron and Tibshirani [1993]. These same authors comment on when and when not to use bias-corrected estimates. Other general texts and articles of interest include Efron and Tibshirani [1986, 1991], Mooney and Duval [1993], Stine [1990] and Young [1994]. DiCiccio and Romano [1988] and Efron and DiCicco [1992] discuss bootstrap confidence by Ducharme et al [1985].

---

[13] A third iteration involving some 800,000,000 numbers will further improve the interval's accuracy.

Techniques for model validation are reviewed in Shao and Tu [1995; p306-313.]. The delete-50% method is shown to be far superior to delete-1.

The smoothed bootstrap was introduced by Efron [1979] and Silverman [1981], and is considered at length in Silverman and Young [1987], Praska Rao [1983], and Falk and Reiss [1989]. For discrete distributions, the approach of Lahiri [1993] is recommended. The weighted bootstrap is analyzed by Barbe and Bertail [1995] and Gleason [1988].

The early literature relied erroneously on estimates of the mean and covariance derived from the original rather than the bootstrap sample. The improved method used today is due to Hall and Wilson [1991]. See guidelines for bootstrap testing in Section 5.3.1.

BC_a intervals and a computationally rapid approximation known as the ABC method are described in Efron [1994]. Bootstrap iteration is introduced in Hall [1986] and Hall and Martin [1988].

Potential flaws in the bootstrap approach are considered by Schenker [1985], Efron [1988, 1992], Knight [1989], and Gine and Zinn [1989].

5.11. Exercises

1. Which sex has longer teeth? Here are the mandible lengths of nine golden jackals (in mm): Males: 107, 110, 116, 114, 113; Females:110, 111, 111, 108. Obtain a 90% confidence interval for the difference in tooth length. Apply both the bootstrap and the permutation test and compare the intervals you derive.
2. Exercise 3.8 described the results of an experiment with an antibiotic used to treat Herpes virus type II. Using the logarithm of the viral titre, determine an approximate 90% confidence interval for the treatment effect. (*Hint*: Keep subtracting a constant from the logarithms of the observations on saline controls until you can no longer detect a treatment difference.) Compare the intervals you derive using the bootstrap and the permutation test.
3. Use the binomial formula of Section 2.2.2 to derive the values of p depicted in Figure 5.2. Here is a time-saving hint: The probability of k or more successes in n trials is the same as the probability of no more than $n - k$ failures.
4. In Table 3.1, breaks appear to be related to log dose in accordance with the formula breaks $= a + b. \log[\text{dose} + .01] + \epsilon$ where ϵ is a random variable with zero mean and a symmetric distribution.

   a. Estimate b.
   b. Use both permutations and the bootstrap to derive confidence intervals for b.
   c. What does a represent? How would you estimate a?

5. What is the correlation between undergraduate GPA and the LSAT? Use the law school data in Exercise 1.14 and the bootstrap to derive a confidence interval for the correlation between them.

6. Suppose you wanted to estimate the population variance. Most people would recommend you use the formula $\frac{1}{n-1}\sum_{i=1}^{n}(x_i - \bar{x})^2$ while others would say, too much work, just use the plug-in estimate $\frac{1}{n}\sum_{i=1}^{n}(x_i - \bar{x})^2$. What do you suppose the difference is? (Need help deciding? Try this experiment: Take successive bootstrap samples of size 11 from the 5th-grade height data in Chapter 1 and use them to estimate the variance of the entire class.)

7. "Use the sample median to estimate the population median and the sample mean to estimate the population mean." This sounds like good advice, but is it? Use the technique described in Exercise 6 to check it out.

8. Obtain smoothed bootstrap confidence intervals for a) the median of the student height data, b) correlation coefficient for the law school data.

9. Can noise affect performance on the job? The following table summarizes the results of a time and motion study conducted in your factory.

Noise (db)	Duration (min)	Noise (db)	Duration (nt)
0	11.5	40	16.0
0	9.5	60	20.0
20	12.5	60	19.5
20	13.0	80	28.5
40	15.0	80	27.5

Draw a graph of duration as a function of noise. Use Pitman correlation to determine if the relationship is significant. Use the bootstrap to obtain a confidence interval for the correlation.

10. Should your tax money be used to fund public television? When a random sample of adults were asked for their responses on a 9-point scale (1 is very favorable and 9 is totally opposed) the results were as follows:

3, 4, 6, 2, 1, 1, 5, 7, 4, 3, 8, 7, 6, 9, 5

Provide a point estimate and confidence interval for the mean response.

11. More than 200 widgets per hour roll off the assembly line to be packed 60 to a carton. Each day you set aside ten of these cartons to be opened and inspected. Today's results were 3, 2, 3, 1, 2, 1, 2, 1, 3, and 5 defectives.

   a. Estimate the mean number of defectives per carton and provide an 80% confidence interval for the mean.

   b. Can you be 90% certain the proportion of defectives is less than 5%?

   c. How does this exercise differ from Exercise 3? (*Hint*: See Section 4.2.)

12. Find a 98% confidence interval for the difference in test scores observed in Exercise 3.12.

   a. Does this interval include zero?

   b. If the new teaching method does not represent a substantial improvement, would you expect the confidence interval to include zero?

13. Compute and compare the apparent error rates of the two Fawlty Towers prediction models.

14. Use the delete-50% method to compare the two Fawlty Towers prediction models.

15. Apply the bootstrap to the billing data provided in Exercise 1.16 to estimate the standard error of the mean billing for each of hospitals One and Two. Compare the ratio of the standard errors with the ratio of the standard deviations for the two hospitals. Why are these ratios so different? Check your intuition by computing the standard error and standard deviation for hospital Three.

CHAPTER 6

Power of a Test

In Chapter 3, you were introduced to some practical, easily computed tests of hypotheses. But are they the best tests one can use? And are they always appropriate? In this chapter, we consider the assumptions that underlie a statistical test and look at some of a test's formal properties: its significance level, power, and robustness.

In our example of the missing labels in Chapter 3, we introduced a statistical test based on the random assignment of labels to treatments, a permutation test. We showed that this test provided a significance level of 5 %, an *exact* significance level, not an approximation. The test we derived is valid under very broad assumptions. The data could have been drawn from a normal distribution (see Figure 4.2) or they could have come from some quite different distribution like the exponential (Figure 4.1). All that is required for our permutation test to be valid is that under the null hypothesis the distribution from which the data in the treatment group is drawn be the same as that from which the untreated sample is taken.

This freedom from reliance on numerous assumptions is a big plus. The fewer the assumptions, the fewer the limitations and the broader the potential applications of a test. Before statisticians introduce a test into their practice, they need to know a few more things about it: How powerful a test is it? That is, how likely is it to pick up actual differences between treated and untreated populations? Is this test as powerful as or more powerful than the test they are using currently? How robust is the new test? That is, how sensitive is it to violations of the underlying assumptions and conditions of an experiment? What if data is missing, as it is in so many practical experiments we perform? Will missing data affect the significance level? What are the effects of extreme values, or *outliers*? Can we extend the method to other, more complex experimental designs in which there are several treatments at several different levels and several simultaneous observations on each subject?

The balance of this chapter provides a theoretical basis for the answers.

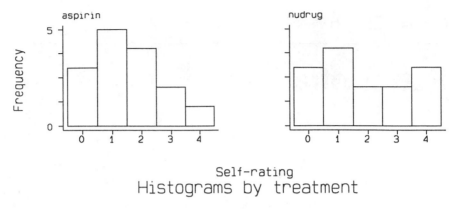

FIGURE 6.1. Patient Self-Rating in Response to Treatment.

6.1.   Fundamental Concepts

In this section, you will be introduced in an informal way to the fundamental concepts of Type I and Type II errors, significance level, power, and exact and unbiased tests.

6.1.1.   Two Types of Error

Figure 6.1 depicts the results of an experiment in which two groups were each given a pain killer. The first group got buffered aspirin; the second group received a new experimental drug. Each participant provided a subjective rating of the effects of the drug. The ratings ranged from "Got worse," to "Much improved"; they are depicted below on a scale from 0 to 5. Take a close look at Figure 6.1. Does the new drug represent an improvement over aspirin?

Some of those who took the new experimental drug do seem to have done better, but not everyone. Are the differences we observe in Figure 6.1 simply the result of chance? Or do they represent a true treatment effect? If it's just a chance effect and we opt in favor of the new drug, we've made an error. We also make an error if we decide there is no difference and the new drug really is better. These decisions and the effects of making them are summarized in Table 6.1.

We distinguish the two types of error because they have the quite different implications described in Table 6.1. As a second example, Fears, Tarone, and Chu [1977] use permutation methods to assess several standard screens for carcinogenicity. Their Type I error, a false positive, consists of labeling a relatively innocuous compound as carcinogenic. Such an action means economic loss for the manufacturer and the denial of the compound's benefits to the public, neither consequence is desirable. But a false negative, a Type II error, would mean exposing a large number of people to a potentially lethal compound.

Because variation is inherent in nature, we are bound to make errors when we draw inferences from experiments and surveys, particularly if chance hands us a

TABLE 6.1. Decision Making under Uncertainty

Our Decision

The Facts	No difference	Drug is Better
	No Difference	Type I error
		Manufacturer misses opportunity for profit
		Public denied access to effective treatment
Drug is better	Type II error	
	Manufacturer wastes money developing ineffective drug	

TABLE 6.2. Decision Making Under Uncertainty

Fears et al's Decision

The Facts		
No effect	Not a Carcinogen	Compound a Carcinogen
		Type I error
		Manufacturer misses opportunity for profit
		Public denied access to effective treatment
Carcinogen	Type II error	
	Patients die;	
	families suffer	
	manufacturer sued	

completely unrepresentative sample. When I toss a coin in the air six times, I can get three heads and three tails, but I can also get six heads. This latter event is less probable, but it is not impossible. Does the best team always win?

We can't eliminate risk in decision making, but we can contain it by choosing the correct statistical procedure. For example, we can require the probability of making a Type I error not exceed 5% (or 1% or 10%) and restrict our choice to statistical methods that ensure we do not exceed this level. If we have a choice of several statistical procedures all of which restrict the Type I error appropriately, we can choose the method that leads to the smallest probability of making a Type II error.

6.1.2. Losses

The preceding discussion is greatly oversimplified. Obviously, our losses will depend not merely on whether we guess right or wrong, but on how far our guesstimate is off the mark. For example, you've developed a new drug to relieve anxiety and are investigating its side effects. Does it raise blood pressure? You do a study and find the answer is no. Alas, the truth is that your drug raises systolic blood pressure an average of 1 mm. What is the cost to the average patient? Not much, negligible.

Now suppose your new drug actually raises blood pressure an average of 10 mm. What is the cost to the average patient? To the entire potential patient population? To your company in lawsuits? One thing is sure, the cost of a type II error depends on the magnitude of that error.

6.1.3.   Significance Level and Power

In selecting a statistical method, statisticians work with two closely related concepts: significance level and power. The *significance level* of a test, denoted throughout the text by the Greek α, is the probability of making a Type I error; that is, α is the probability of deciding erroneously on the alternative when the hypothesis is true.

To test a hypothesis, we divide the set of possible outcomes into two or more regions. We accept the hypothesis and risk a Type II error when our test statistic lies in the *acceptance region*, A; we reject the hypothesis and risk a Type I error when our test statistic lies in the *rejection region*, R; and we may take additional observations when our test statistic lies in the boundary region of *indifference*, I. If H denotes the hypothesis, then

$$\alpha = \Pr\{A|H\}.$$

The *power* of a test, denoted throughout the text by the Greek β, is the complement of the probability of making a Type II error; that is, β is the probability of deciding on the hypothesis when the alternative is the correct choice. If K denotes the alternative, then

$$\beta = \Pr\{R|K\}.$$

The ideal statistical test would have a significance level α of zero, or 0%, and a power β of 1, or 100%. But unless we are all-knowing, this ideal cannot be realized. In practice, we fix a significance level $\alpha > 0$, where α is the largest value we feel comfortable with, and choose a statistic that maximizes or comes closest to maximizing the power.

Power and Sample Size

As we saw in Section 6.1.2, the greater the discrepancy between the true alternative and our hypothesis, the greater the loss associated with a Type II error. Fortunately, in most practical situations, we can devise a test where the larger the discrepancy, the greater the power and the less likely we are to make a Type II error.

Figure 6.2 depicts the power as a function of the alternative for two tests based on samples of size 6. In the example illustrated, the test ϕ_1 is uniformly more powerful than ϕ_2; hence, using ϕ_1 in preference to ϕ_2 will expose us to less risk.

Figure 6.3 depicts the power curve of these same two tests using different size samples; the power curve of ϕ_1 is still based on a sample of size 6, but that of ϕ_2 now is based on a sample of size 12. The two new power curves almost coincide, revealing the two tests now have equal risks. But we will have to pay for twice as many observations if we use the second test in place of the first.

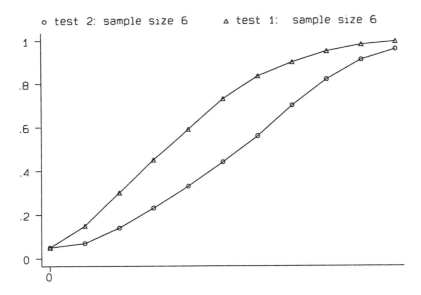

FIGURE 6.2. Power as a Function of the Alternative; Both Tests Have the Same Sample Size.

Moral: *A more powerful test reduces the costs of experimentation along with minimizing the risk.*[1]

Power and the Alternative

If a test at a specific significance level α is more powerful against a specific alternative than all other tests at the same significance level, we term it *most powerful*. But as we see in Figure 6.4, a test that is most powerful for some alternatives may be less powerful for others. When a test at a specific significance level is more powerful against all alternatives than all other tests at the same significance level, we term this test uniformly most powerful.

The significance level and power may also depend on how the variables we observe are distributed. Does the population distribution follow a bell-shaped normal curve, with the most frequent values in the center? Or is the distribution something quite different? To protect our interests, we may need to require that the Type I error be less than or equal to some predetermined value for all possible distributions. When applied correctly, permutation tests always have this property. The significance level of a test based on the bootstrap is dependent on the underlying distribution.

---

[1]The exception proves the rule. When data gathering is dirt cheap, Lloyd Nelson observes, a less powerful test, such as the sign test, makes economic sense.

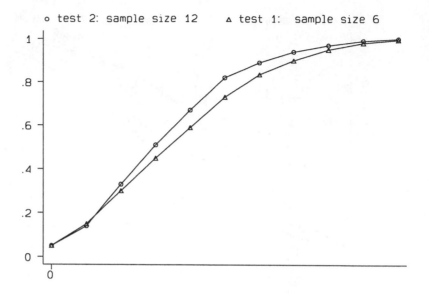

FIGURE 6.3. Power as a Function of the Alternative; Tests Have Different Sample Sizes.

The *power* of a test depends upon the statistic, the sample size, and the alternative.

6.1.4.　Exact, Unbiased Tests

In practice, we seldom know the distribution of a variable or its variance. Suppose we wish to test the hypothesis "X has mean 0." This *compound hypothesis* includes several simple hypotheses such as H_1: X is normal with mean 0 and variance 1; H_2: X is normal with mean 0 and variance 1.2, and H_3: X is a gamma distribution with mean 0 and four degrees of freedom.

A test is said to be *exact* with respect to a compound hypothesis if the probability of making a Type I error is exactly α for each and every one of the possibilities that make up the hypothesis. A test is said to be *conservative* if the Type I error never exceeds α. Obviously, an exact test is conservative, but the reverse may not be true.

The importance of an exact test cannot be overestimated, particularly a test that is exact regardless of the underlying distribution. If a test that is nominally at level α is actually at level γ, we may be in trouble before we start. If $\gamma > \alpha$, the risk of a Type I error is greater than we are willing to bear. If $\gamma < \alpha$, then our test is suboptimal, and we can improve it by enlarging its rejection region. We'll return to these points again in Chapter 12 on choosing a statistical method.

A test is said to be *unbiased* and of level α, provided its power function β satisfies the following two conditions:

1. β is conservative; that is, $\beta(\theta) \leq \alpha$ for every θ that satisfies the hypothesis.
2. $\beta(\theta) \geq \alpha$ for every θ that is an alternative to the hypothesis.

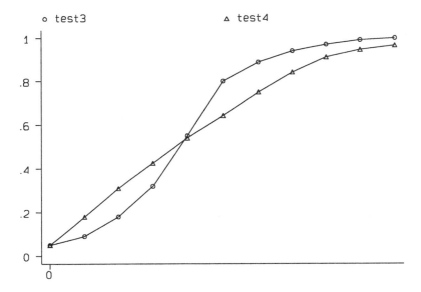

FIGURE 6.4. Comparing Power Curves: For Near Alternatives, with θ Close to Zero, Test 4 Is the More Powerful Test; for Far Alternatives with θ Large, Test 3 Is More Powerful. Thus Neither Test is Uniformly Most Powerful.

In other words, a test is unbiased if it is more likely to reject a false hypothesis than a true one.[2] If $\{A(\theta)\}$ is a collection of acceptance regions associated with a uniformly most powerful unbiased test, the correct value of the parameter is more likely to be covered by the confidence intervals we construct from $\{A(\theta)\}$ than an incorrect value.

Faced with some new experimental situation, our objective is always to derive a uniformly most powerful unbiased test if one exists. But, if we can't derive a uniformly most powerful test, and if Figure 6.4 depicts just such a situation, then we will look for a test that is most powerful against those alternatives that are of immediate interest.

6.1.5. Dollars and Decisions

We now know that a statistical problem is defined by four elements:

1. The observations $X = \{X_1, X_2, \cdots X_n\}$.
2. The distribution F of the variables $\{X_i\}$ in the population from which the observations are drawn.
3. The set $\{d\}$ of possible decisions one can make on observing X.
4. The loss $L(d, \theta)$, expressed in dollars, men's lives, or some other quantifiable measure, that results when we make the decision d when θ is true.

---

[2] I find unbiasedness to be a natural and desirable principle, but not everyone shares this view; see, for example, Suissa and Shuster [1984].

A problem is statistical when the investigator is not in a position to say that X will take on exactly the value x, but only that X has some probability $P\{x \in A\}$ of taking on values in the set A.

In this text, we've limited ourselves for the most part to two-sided decisions in which either we accept a hypothesis H and reject an alternative K, or we reject the hypothesis H and accept the alternative K. One example is H: Population mean $\leq \theta_o$ vs K: Population mean $\geq \theta_o$. As in Section 5.2, we would probably follow our decision to accept or reject with a confidence interval for the unknown parameter, such as $\theta_1 <$ Mean, θ_2, and a statement to the effect the probability this interval covers the true parameter value is not less than $1 - \alpha$.

Typically, losses L depend on some function of the difference between the true (but unknown) value of the population mean θ and our best guess (point estimate) of its value $\theta*$; $L(\theta, \theta*) = |\theta - \theta*|$, for example[3], or $L(\theta, \theta*) = (\theta - \theta*)^2$. Our objective is to come up with a decision rule D, such that when we average over all possible samples we might draw from the population F, we minimize the associated *risk*

$$R_\theta(D) = \int L[D(X)]dF_\theta \text{ or } \sum L(D[x_1, x_2, \dots, x_n]) \Pr(x_1, x_2, \dots, x_n | \theta).$$

Unfortunately, a testing procedure that is optimal for one value of the parameter θ might not be optimal for another. This situation was illustrated in Figure 6.4, with two decision curves that cross each other. My tentative solution to this complex problem is to always use an unbiased procedure.[4]

6.1.6. What Significance Level Should I Use?

Choice of significance level and power are determined by the environment in which you work. Most scientists simply report the observed p-value, leaving it to their peers to decide for themselves what weight should be given to the results.

Significance Level, Power, and Sample Size

Significance level, power, and sample size are interelated. Fix the significance level at a given sample size and you determine the power against all alternatives. Increase the sample size without changing the significance level and you increase the power. If you are willing to accept a greater risk of making a Type I error, then you can take fewer observations without affecting the power. Of course, both power and significance level depend on your choice of test statistic.

---

[3] A practical example is given in Section 4.3.1.

[4] This problem is complex with philosophical as well as mathematical overtones; we refer the interested reader to the discussions in the first chapter of Erich Lehmann's book, *Testing Statistical Hypotheses* [1986].

A manufacturer preparing to launch a new product line or a pharmaceutical company conducting research for promising compounds typically adopts a three-way decision procedure: If the observed p-value is less than 1%, it goes forward with the project. If the p-value is greater than 20%, it abandons it. And if the p-value lies in the gray area in between, it arranges for additional surveys or experiments.

A regulatory commission like the FDA which is charged with oversight responsibility, must work at a fixed significance level, typically $\alpha = 0.05$ or 0.10. The choice of a fixed significance level ensures consistency in both result and interpretation as the agency reviews the findings from literally thousands of tests. They specify a minimum power level of 90% or 95%, or even 99% against biologically significant alternatives if potentially dangerous side effects are under investigation.

6.2. Assumptions

Every test carries with it certain assumptions, for example, that the sample was selected at random from the population. These assumptions must be satisfied in order to achieve a desired significance level.

Many of the statistical procedures in common use (see Chapter 4) rely on the distribution of the underlying data having a specific form, such as the normal, binomial, or Poisson. If this assumption is correct, *parametric* statistical procedures such as the t-test or the F-test may be called for. If the assumption is in error, parametric procedures may yield erroneous results.

Underlying the bootstrap are two assumptions: (a) if the null hypothesis is true then all observations in the sample (or subsample) come from populations that have the same value of the parameter being estimated and (b) the observations are independent of one another.

Underlying the permutation test is the assumption that if the null hypothesis is true all observations in the sample (or subsample) are *exchangeable*. Loosely speaking, observations are exchangeable if (a) they are independent and identically distributed or (b) they are identically normally distributed and have the same mutual correlations, or (c) they are obtained by sampling without replacement from the same finite population. In the one-sample case, as well as matched pairs (and regression), the permutation test requires the further assumption that the observations (or residuals) come from a symmetric distribution.

Sometimes a simple transformation will ensure observations are exchangeable or, in the case of the bootstrap, have the same parameter. For example, if we know that X comes from a population with mean μ and distribution $F(x - \mu)$ and an independent observation Y comes from a population with mean v and distribution $F(x - v)$, then the independent variables $X' = X - \mu$ and $Y' = Y - v$ are exchangeable.

The key issue with permutation tests is whether under the null hypothesis of no differences among the various experimental or survey groups, we can exchange

---

Which Test?

If you can exchange the labels on the observations without affecting the results, it is safe to use a *permutation test*.

If the observations have the same population parameters when the null hypothesis is true, it is safe to use the *bootstrap*.

---

the labels on the observations without affecting the results. If we have taken steps to ensure our sampling method is representative (as described in Section 2.6) and performed any necessary equalizing transformations (see Sections 3.3.1, 3.4, and 3.8), then the answer in almost all cases will be a resounding "*yes.*"

6.3.   How Powerful Are Our Tests?

Given that these assumptions are satisfied, how effective and powerful are the tests we introduced in preceding chapters?

6.3.1.   One-Sample

Our permutation test is applicable even if the different observations come from different distributions, provided these distributions are all symmetric and all have the same location parameter or median.[5] Against specific normal alternatives, our permutation test provides a most powerful unbiased test of the distribution-free hypothesis H: $\theta = \theta_o$ [Lehmann, 1986, p. 239]. Even if we know that the underlying data has the normal distribution (and it is rare that we would know), for large samples, the permutation test's power is almost the same as the most powerful parametric test, Student's t (Albers, Bickel, and van Zwet, 1976). Even if the underlying distributions are almost (but not quite) symmetric, Romano [1990] shows that for very large samples, the one-sample permutation test for a location parameter is exact, provided the underlying distribution has finite variance. His result applies whether the permutation test is based on the mean, the median, or some statistical function of the location parameter. If the underlying distribution is almost symmetric, the permutation test will be almost exact even when based on as few as 10 or 12 observations.

The bootstrap is neither exact nor conservative, but if the observations are independent and from distributions with identical values of the parameter of interest, the bootstrap is exact for very large samples [Liu, 1988].[6] A nonparametric boot-

---

[5] A simple transformation can often ensure that this requirement is satisfied; for example, subtracting a constant from all members of the second sample or working with logarithms rather than the original observations. See Section 3.8.

[6] By "very large," we mean 1,000, or 10,000, or even more observations.

Guidelines for Bootstrap Hypothesis Testing

The bootstrap is neither exact nor conservative and may have very low power even against distant alternatives. To increase power and improve accuracy, Hall and Wilson [1991] provide the following guidelines for a test of H: $\theta = \theta_o$:

1. Resample $\widehat{\theta}^* - \widehat{\theta}$ not $\widehat{\theta}^* - \theta_o$, where $\widehat{\theta}$ is the estimate of θ obtained from the original sample, and $\widehat{\theta}^*$ is the estimate obtained from the bootstrap sample.
2. Base the test on the bootstrap distribution of $(\widehat{\theta}^* - \widehat{\theta})/\widehat{\sigma}^*$, where $\widehat{\sigma}^*$ is an estimate of the standard deviation of $\widehat{\theta}$ based on the bootstrap sample, not on the bootstrap distribution of $(\widehat{\theta}^* - \widehat{\theta})/\widehat{\sigma}$ or $\widehat{\theta}^* - \widehat{\theta}$.

---

strap is less powerful than a permutation test, but in the one-sample case does not require the underlying distribution to be symmetric.[7]

6.3.2. Matched Pairs

The results for a single sample apply here as well, with the added bonus that we know the underlying data is symmetric (see, for example, Good, 1994, p. 41).

6.3.3. Regression

An initial tranformation converts the regression problem studied in Section 5.3 to that of the single-sample method. To test that $y = Ax$ where A is a specific matrix of constants, we apply the resampling methods to the residuals $y - Ax$. Our assumptions and findings are the same as in the one-sample case.

6.3.4. Two Samples

A permutation test based on our statistic is exact and unbiased against stochastically increasing alternatives of the form $K : F_1[x] = F[x - d]$ for $d > 0$. In fact, this permutation test is a uniformly most powerful unbiased test of the null hypothesis $H : F_1 = F$ against normally distributed shift alternatives. Against normal alternatives and for large samples, its power is equal to that of the standard t-test described in Section 4.3.2 [Bickel and van Zwet, 1978].

The permutation test offers the advantage over the parametric t-test that it is exact even for very small samples *whether or not the observations come from a normal distribution*. The parametric t-test relies on the existence of a mythical infinite population from which all the observations are drawn. The permutation test is applicable even to finite populations such as all the machines in a given shop or all the supercomputers in the world.

---

[7]As noted in Chapter 4, when the underlying distribution is not symmetric, a sample with at least 50 observations is essential if the boostrap is to be a valid approximation.

The bootstrap does not require that the observations all come from the same population, only that they all have the same value of the population parameter under investigation. The bootstrap is neither exact nor conservative, but if the observations are independent and from distributions with identical values of the parameter of interest, then the bootstrap is exact for very large samples [Liu, 1988]. The number of observations required for an almost-exact bootstrap can be reduced by using BC_a intervals described in Section 5.7 and the Hall-Wilson guidelines outlined in the accompanying sidebar.

A nonparametric bootstrap is less powerful than a permutation test. Still, the bootstrap may be applicable when neither a permutation test nor a parametric test exists.

6.4.   Which Test?

We can now make an initial comparison of the three types of statistical tests— permutation, bootstrap, and parametric.[8]

Recall from Chapter 3 that with a permutation test, we:

1. Choose a test statistic $S(X)$.
2. Compute S for the original set of observations.
3. Obtain the permutation distribution of S by repeatedly rearranging the observations. With two or more samples, we combine all the observations into a single large sample before we rearrange them.
4. Obtain the upper α-percentage point of the permutation distribution and accept or reject the null hypothesis according to whether S for the original observations is smaller or larger than this value.

If the observations are exchangeable, then the resultant test is exact and unbiased.

The nonparametric bootstrap, like the permutation test, requires a minimum number of assumptions and derives its critical values from the data at hand.

Recall from Chapter 5 that to obtain a nonparametric bootstrap, we:

1. Choose a test statistic $S(X)$.
2. Compute S for the original set of observations.
3. Obtain the bootstrap distribution of S by repeatedly resampling with replacement from the observations. We need not combine the samples but may resample separately from each sample.
4. Smooth the bootstrap distribution and/or apply bias corrections.
5. Obtain the upper α-percentage point of the corrected bootstrap distribution and accept or reject the null hypothesis according to whether S for the original observations is smaller or larger than this value.

Recall from Chapter 5 that to obtain a parametric test (e.g., a t-test or an F-test), we:

---

[8] A fourth type, tests based on density estimation, will be considered in Chapter 8.

1. Choose a test statistic S whose distribution to F_S may be computed and tabulated independent of the observations.
2. Compute S for the observations X.
3. Compare $S(X)$ with the upper α-percentage point of F_S and accept or reject the null hypothesis according to whether $S(X)$ is smaller or larger than this value.

If the observations are independent, the sample selected by representative means, and S is distributed as F_S, then the parametric test is exact and, often, the most powerful test available. If S really has some other distribution, then the parametric test may lack power and may not be conservative. Singh [1981] was the first to demonstrate the advantages of the bootstrap over large-sample approximations. With large[9] samples, the permutation test is usually as powerful as the most powerful parametric test [Bickel and Van Zwet, 1978]. If S is not distributed as F_S, it may be more powerful.

6.5.   Summary

In this chapter, you learned that power, sample size, and significance level are interrelated. You learned the value of exact unbiased tests. You learned that your choice of test statistic will depend on the hypothesis, the alternative, the loss function, and the type of test you're using. You learned the assumptions underlying and the differences among test of hypotheses based on the bootstrap, the permutation test, and parametric distributions.

6.6.   To Learn More

A formal discussion of risk theory is found in Lehmann [1986]. For a further discussion of exchangeability, see Lehmann [1986, p. 231]; Koch [1982]; and Draper et al [1933]. Examples of power calculations for resampling methods are given in Oden [1975]. Keller-McNulty and Higgens [1987] and Hall and Titterington [1989]. For a discussion of the relative advantages of variable versus fixed significance levels, see Kempthorne [1966].

6.7.   Exercises

1. **a.** Sketch the power curve $\beta(\theta)$ for one of the two-sample comparisons described in Chapter 3. (You already know two of the values for each power curve; what are they?)

---

[9]How large is *large*? If we are investigating a location parameter, it could be as few as 10 or 12 observations; in other instances, involving ratios of several parameters, it could mean 100 or more.

    b. Using the same set of axes, sketch the power curve of a test based on a much larger sample.

    c. Suppose that without looking at the data you: (i) always reject; (ii) always accept; or (iii) use a chance device so as to reject with probability α. For each of these three tests, determine the power and the significance level. Are any of these three tests exact? Unbiased?

2. True or False? Tests should be designed so that:

    a. The risk of making a Type I error is low.

    b. The probability of making a Type II error is high.

    c. The null hypothesis is likely to be rejected.

    d. Economic consequences of a decision are considered.

3. True or False? The expected loss from using a particular statistical procedure will depend on:

    a. The probability that the procedure will lead to a wrong decision.

    b. The losses associated with a wrong decision.

4. **a.** Suppose you have two potentially different radioactive isotopes with half-life parameters λ_1 and λ_2, respectively. You gather data on the two isotopes and, taking advantage of a uniformly most powerful unbiased permutation test, you reject the null hypothesis H: $\lambda_1 = \lambda_2$ in favor of the one-sided alternative K:$\lambda_1 > \lambda_2$. What are you or the person you are advising going to do about it? Will you need an estimate of λ_1/λ_2? What estimate will you use?

    b. Review some of the hypotheses you tested in the past. Distinguish your actions after the test was performed from the conclusions you reached. (In other words, did you do more testing? Rush to publication? Abandon a promising line of research?) What losses were connected with your actions? Should you have used a higher/lower significance level? Should you have used a more powerful test or taken more or fewer observations? And, if you used a parametric test like Student's t or Welch's z, were all the assumptions for these tests satisfied?

5. Late for a date, you dash out of the house, slipping a bill into your wallet. Was it a $2 bill or a $20? Too late now. At the movie theater, your friend slips a $2 bill into your hand, "for my share." (A share of what? The popcorn?) You put this bill away in your wallet too. As you approach the ticket window, you decide to perform a simple statistical test. You'll take a quick look at one of the bills. If it's a $2 bill, you'll accept the hypothesis that both bills are $2. Otherwise, you'll reject this hypothesis. What is the significance level of your test? The power?

6. Given a choice of a permutation test based on the original scores, a permutation test based on ranks, a bootstrap, and the best and most appropriate parametric test, which would you use in the following situations and why? [*Hint*: You may want to glance through Chapters 3, 4, and 5 a second time.] In cases a–e, you want to test a hypothesis about the median of a population:

a. You have 10 observations from a normal distribution.

b. You have 10 observations from an almost-normal distribution.

c. You have 10 observations from an exponential distribution.

d. You have 10 observations from an almost-exponential distribution.

e. You have 50 observations from an almost-exponential distribution.

In cases f–h, you want to compare two populations:

f. You have two samples of size 10 taken from normal distributions with the same variance and want to compare medians.

g. You have two samples of size 10 taken from similarly shaped distributions and want to compare medians. One problem: All the data is in the hundreds, except for one observation which is 10.1. Or is that decimal point a flaw in the paper?

h. You have two samples of size 25 and want to compare variances.

7. a. Your lab has been presented with a new instrument offering 10 times the precision of your present model. How might this affect the power of your tests? Their significance level? The number of samples you'll need to take?

b. A directive from above has loosened the purse strings so you now can take larger samples. How might this affect the power of your tests? Their significance level? The precision of your observations? The precision of your results?

c. A series of law suits over those silicon implants you thought were harmless has totally changed your company's point of view. How might this affect the power of your tests? Their significance level? The precision of your observations? The number of samples you'll take? The precision of your results?

8. Turn to the literature of your field and consider some of the estimates that have been made. Which do you feel is the most appropriate loss function in each case $L_1 = |\theta - \theta*|$ or $L_2 = (\theta - \theta*)^2$? If hypotheses were tested, were all test assumptions satisfied?

9. Imagine your company, a manufacturer of industrial supplies, plans to launch a new product to sell to existing customers. Development costs would be approximately $1 million. Your anticipated profit is $1,000 per unit sold. You have a customer base of 10,000. What percentage would you need to sell to break even? Of course, the actual percentage you sell might be greater or less than this amount. If you decide to manufacture the product, what sort of losses might you experience? If you decide against development, what profits might you be surrendering?

You decide to do a survey of your customers and use this survey to estimate the percentage you will sell. If p is your estimate and π is the actual value of this percentage, construct a graph of losses in dollars versus the estimate error $(\pi - p)$.

10. Your company policy is to insist on a 99% compliance policy in the components you purchase. That is, no more than 1% of the units can be out of compliance. How many units would you need to examine to be sure of rejecting a lot shipped to you with 2% defective components? Examining all of them would get the job done, but there must be a less expensive way. Make a chart for $n = 100, 200,$ and so on, in which you list k (the maximum number of tested units that can be out of compliance), α (the resulting significance level when 99% of the components are actually in compliance), and β (the power when only 98% of the units are actually in compliance).

11. Take a second look at the experimental methods described in Exercises 3.12 and 3.13. Which approach should yield the more powerful test?

12. Take a second look at the process temperature data of Cox and Snell, [1981] in Table 3.2. What would be the effect on the significance level and power of a two-tailed bootstrap test of the hypothesis $\theta = 440$ if we follow the first of the Hall-Wilson guidelines? The second guideline?

13. Formulate hypotheses and alternatives for comparing the billing practices of the four hospitals considered in Exercise 1.16. What will you compare? Means? Medians? Variances? Frequency distributions? (You may wish to return to this question after completing Chapters 7 and 8.)

CHAPTER 7

Categorical Data

In many experiments and almost all surveys, many, if not all, of the results fall into categories rather than being measurable on a continuous or ordinal scale: e.g., male vs. female, black vs. Hispanic vs. Asian vs. white, in favor vs. against vs. undecided. The corresponding hypotheses concern proportions: "Blacks are as likely to be Democrats as they are to be Republicans." Or, "the dominant genotype 'spotted shell' occurs with three times the frequency of the recessive." In this chapter, you'll learn to test hypotheses like these that concern categorical and ordinal data.

7.1. Fisher's Exact Test

As an example, suppose on examining the cancer registry in a hospital, we uncover the following data, which we put in the form of a 2×2 *contingency table*:

7.1	Survived	Died	Total
Men	9	1	10
Women	4	10	14
	13	11	24

The *9* denotes the number of males who survived, the *1* denotes the number of males who died, and so forth. The four marginal totals, or *marginals*, are 10, 14, 13, and 11. The total number of men in the study is 10, 14 denotes the total number of women, and so on.

We see in this table an apparent difference in the survival rates for men and women: Only 1 of 10 men died following treatment, but 10 of the 14 women failed to survive. Is this difference statistically significant?

The answer is *yes*. Let's see why, using the same line of reasoning that Fisher advanced at the annual Christmas meeting of the Royal Statistical Society in 1934. After Fisher's talk concluded, incidentally, speaker compared Fisher's talk to "the braying of the Golden Ass." I hope you will take more kindly to my own explanation. The preceding contingency table has several fixed elements: the total number of men in the survey, 10, the total number of women, 14, the total number who died, 11, and the total number who survived, 13. These totals are immutable; no swapping of labels will alter the total number of individual men and women or bring back the dead. But these totals do not determine the content of the table, as can be seen from the following two tables whose marginals are identical with those of our original table.

7.2	Survived	Died	Total
Men	10	0	10
Women	3	11	14
	13	11	24

7.3	Survived	Died	Total
Men	8	2	10
Women	5	9	14
	13	11	24

The first of these tables makes a strong case for the superior fitness of the male, stronger even than our original observations. In the second table, the survival rates for men and women are more alike than in our original table.

Fisher would argue that if the survival rates were the same for both sexes, then each of the redistributions of labels to subjects, that is, each of the N possible contingency tables with these same four fixed marginals, is equally likely, where[1]

$$N = \sum_{x=0}^{10} \binom{13}{x}\binom{11}{10-x} = \binom{13+11}{10}.$$

How did we get this value for N? The component terms are taken from the hypergeometric distribution:

$$\sum_{x=0}^{t} \binom{m}{x}\binom{n}{t-x} / \binom{m+n}{t} \tag{7.1}$$

where n, m, t, and x occur as the indicated elements in the following 2×2 contingency table

---

[1]Combinatorial notation, such as choose x of 13 things, was defined in Section 2.2.2.1.

	Category 1	Category 2	
Category A	x	$t - x$	t
Category B	$m - x$	$n - (t - x)$	
	m	n	$m + n$

If men and women have the same probability of surviving, then all tables with the marginals m, n, and t are equally likely, and $\sum_{k=0}^{t-x} \binom{m}{t-k}\binom{n}{k}$ are as extreme or more extreme.

In our example, $m = 13$, $n = 11$, $x = 9$, and $t = 10$, so that $\binom{13}{9}\binom{11}{1}$ of the N tables are as extreme as our original table and $\binom{13}{10}\binom{11}{0}$ are more extreme.

$11\binom{13}{9} + \binom{13}{10}$ is a very small fraction of the total, so we conclude that a difference in survival rates as extreme as the difference we observed in our original table is very unlikely to have occurred by chance. We reject the hypothesis that the survival rates for the two sexes are the same and accept the alternative that, in this case at least, males are more likely to profit from treatment.

7.1.1.   One-Tailed and Two-Tailed Tests

In the preceding example, we tested the hypothesis that survival rates do not depend on sex against the alternative that men diagnosed as having cancer are likely to live longer than women similarly diagnosed. We rejected the null hypothesis because only a small fraction of the possible tables were as extreme as the one we observed initially. This is an example of a one-tailed test. Or is it? Wouldn't we have been just as likely to reject the null hypothesis if we had observed a table of the following form:

7.4	Survived	Died	Total
Men	0	10	10
Women	13	1	14
	13	11	24

Of course, we would. In determining the significance level in the current example, we must add together the total number of tables that lie in either of the two extremes or tails of the permutation distribution.

McKinney et al. [1989] reviewed more than 70 articles that appeared in 6 medical journals. In more than half of these articles, Fisher's exact test was applied improperly. Either a one-tailed test had been used when a two-tailed test was called for or the authors of the paper simply hadn't bothered to state which test they had used.

When you design an experiment, decide at the same time whether you wish to test your hypothesis against a two-sided or a one-sided alternative. A two-sided alternative dictates a two-tailed test; a one-sided alternative dictates a one-tailed test.

As an example, suppose we decide to do a follow-on study of the cancer registry to confirm our original finding that men diagnosed as having tumors live significantly longer than women similarly diagnosed. In this follow-on study, we have a one-sided alternative. Thus, we would analyze the results using a one-tailed test rather than the two-tailed test we applied in the original study.

7.1.2.   The Two-Tailed Test

Unfortunately, it is not as obvious which tables should be included in the second tail. Is 7.4 as extreme as 7.2? We need to define a test statistic as a basis of comparison. One commonly used measure is the χ^2 statistic defined for the 2×2 contingency table after eliminating invariants as $(x - t \frac{m}{m+n})^2$. For 7.1, this statistic is 13, for 7.4, it is 29. We leave it to you to do the computations to show that 7.5 is more extreme than 7.1, but 7.6 is not.

7.5	Survived	Died	Total
Men	1	9	10
Women	12	2	14
	13	11	24

7.6	Survived	Died	Total
Men	2	8	10
Women	11	3	14
	13	11	24

7.2.   Odds Ratio

In most instances, we won't be satisfied with merely rejecting the null hypothesis but we'll want to make some more powerful statement like "men are twice as likely as women to get a good-paying job," or "women under 30 are twice as likely as men over 40 to receive an academic appointment."

In the discrimination case of Fisher versus Transco Services of Milwaukee (1992), the plaintiffs claimed that Transco was 10 times more likely to fire older employees. Can we support this claim with statistics? The Transco data are provided in Table 7.1. Let π_1 denote the probability of firing a young person and π_2

TABLE 7.1. Transco Employment

Outcome	Young	Old
Fired	1	10
Retained	24	17

the probability of firing an older person. We want to go beyond testing the null hypothesis $\pi_1 = \pi_2$ to determine a confidence interval for the odds ratio $\frac{\pi_2}{1-\pi_2} / \frac{\pi_1}{1-\pi_1}$. This problem is too complex and there are too many rearrangements to rely on hand calculations. We turn for help to StatXact-3, a statistical package whose emphasis is the analysis of categorical and ordinal data. Choosing **Statistics**, **Two Binomials**, and **CI. Odds Ratio** from successive StatXact-3 menus, we obtain the following results:

```
StatXact-3 Output
Datafile: C:\SX3WIN\EXAMPLES\TRNSCO.CY3

ODDS RATIO OF TWO BINOMIAL PROPORTIONS

Statistic based on the observed 2 by 2 table:
    Binomial proportion for column <young>:pi_1 = 0.04000
    Binomial proportion for column <Old>:
    pi_2 = 0.3704 ( pi_2 )/(1-pi_2 )
    Odds Ratio = -------------------
    = 14.12 ( pi_1 )/(1-pi_1 )
Results:
Method P-value(2-sided) 95.00% Confidence Interval
Asymp (Mantel-Haenszel) 0.0157 (1.649,120.9)
Exact 0.007145 (1.649, 637.5)
```

Based on these results, we can tell the judge that older workers were fired at a rate at least 1.6 times the rate at which younger workers were discharged.

7.2.1.   Stratified 2×2's

In trying to develop a cure for a relatively rare disease, we face the problem of having to gather data from a multitude of test centers, each with its own set of procedures and way of executing them. Before we can combine the data, we must be sure the odds ratios across the test centers are approximately the same. Consider the set of results in Table 7.2 obtained by the Sandoz Drug Company and reproduced with permission from the StatXact-3 manual.

One of the cites, number 15, stands out from the rest. But is the difference statistically significant?[2] With 22 contingency tables, the number of computations needed to examine all rearrangements is in the billions.

Fortunately, StatXact uses several time-saving algorithms, including the one introduced in Mehta, Patel, and Senchaudhuri [1988], to obtain a Monte Carlo

---

[2]Similar problems were encountered in a study where test subjects might use one of several different "identical" machines. I couldn't combine the results from the different machines or the different technicians who operated them until after I performed an initial test of their equivalence.

TABLE 7.2. Sandoz Drug Data

Test Site	New Drug Response	#	Control Drug Response	#
1	0	15	0	15
2	0	39	6	32
3	1	20	3	18
4	1	14	2	15
5	1	20	2	19
6	0	12	2	10
7	3	49	10	42
8	0	19	2	17
9	1	14	0	15
10	2	26	2	27
11	0	19	2	18
12	0	12	1	11
13	0	24	5	19
14	2	10	2	11
15	0	14	11	3
16	0	53	4	48
17	0	20	0	20
18	0	21	0	21
19	1	50	1	48
20	0	13	1	13
21	0	13	1	13
22	0	21	0	21

estimate of the significance level. We pull down menus **Statistics, Stratified** 2×2 **Tables**, and **Homogeneity of Odds Ratios**, to obtain the following results:

```
StatXact-3 Output

Datafile: C:\SX3WIN\EXAMPLES\SANDOZ.CY3

TEST FOR HOMOGENEITY OF ODDS RATIOS
[18 2x2 informative tables]

Observed Statistics:
    BD: Breslow and Day Statistic = 25.78
    ZE: Zelen Statistic = 9.481e-009

Asymptotic p-value: (based on Chi-Square distribution with
        17 df )
    Pr { BD .GE. 25.78 } = 0.0785

Monte Carlo estimate of p-value:
```

Do It Yourself?

You could write a computer program to perform the tests described in this chapter, that would select from the set of all tables with a given set of marginals to generate the permutation distribution, but there's a more efficient way. Branch-and-bound algorithms developed by Mehta and Patel [1980, 1983] use a network approach to enumerate only those tables that have a more extreme value of the test statistic than the original. An outline of their method is given in section 13.4.1.1 of Good [1994]. StatXact is the only version of these algorithms that is commercially available now, and in what follows, you'll learn how to use StatXact to do each of the needed tasks.

```
Pr { ZE .GE. 9.481e-009 } = 0.0127
99.00% Confidence Interval = ( 0.0119, 0.0135)
```

```
Elapsed Time is 0:16:15.37 (10000 tables sampled with
              starting seed 85190)
```

The estimated p-value of .013, just a fraction greater than 1%, tells us it would be unwise to combine the results from the different cites.

The output of this program provides us with one more important finding: Displayed above the Monte Carlo estimate of the exact p-value, 0.01237, is the asymptotic or large-sample approximation based on the chi-square distribution. Its value, 0.0785, is many times larger than the correct value, and relying on this so-called approximation would have led us to a completely different and erroneous conclusion.

7.3. Exact Significance Levels

The preceding result is not an isolated one. Asymptotic approximations are to be avoided except with very large samples. Table 7.3 contains data on oral lesions in three regions of India derived from Gupta et al. [1980] by Mehta and Patel. We want to test the hypothesis that the location of oral lesions is unrelated to geographical region. Possible test statistics include Freeman-Halton p (see Section 7.4), p_χ and p_L. This latter statistic is based on the log-likelihood ratio $\Sigma \Sigma f_{ij} \log(f_{ij} f_{..} / f_{i.} f_{.j})$.

We may calculate the exact significance levels of these test statistics by deriving their permutation distributions or use asymptotic approximations obtained from tables of the chi-square statistic. Table 7.4 taken from the StatXact-3 manual compares the various approaches.

The exact significance level varies from 1% to 3.5%, depending on which test statistic we select. Tabulated p-values based on large-sample approximations vary from 11% to 23%. Using the Freeman-Halton statistic, the permutation test tells us that differences among regions are significant at the 1% level; the large-sample ap-

TABLE 7.3. Oral Lesions in Three Regions of India

Site of Lesion	Kerala	Gujarat	Andh
Labial Mucosa	0	1	0
Buccal Mucosa	8	1	8
Commissure	0	1	0
Gingiva	0	1	0
Hard Palate	0	1	0
Soft Palate	0	1	0
Tongue	0	1	0
Floor of Mouth	1	0	1
Alveolar Ridge	1	0	1

TABLE 7.4. Three Tests of Independence

Statistic	χ^2	$F - H$	LR
Exact p-value	.0269	.0101	.0356
Tabulated p-value	.1400	.2331	.1060

proximation says no, they are insignificant even at the 20% level. The permutation test is correct. The large-sample approximation is grossly in error. With so many near-zero entries in the original contingency table, the chi-square large-sample approximation is not appropriate.

7.4.   Unordered $r \times c$ Contingency Tables

With a computer at hand, the principal issue in the analysis of a contingency table with r rows ($r > 2$) and c columns ($c > 2$) is deciding on an appropriate test statistic. Halter [1969] showed we can find the probabilities of any individual $r \times c$ contingency table through a straightforward generalization of the hypergeometric distribution given in Equation 7.1. An $r \times c$ contingency table consists of a set of frequencies

$$\{f_{ij}, 1 \le i \le r; 1 \le j \le c\}$$

with row marginals $\{f_{i.}, 1 \le i \le r\}$ and column marginals $\{f_{.j}, 1 \le j \le c\}$. Suppose once again we have mixed up the labels. To make matters worse, this time every item/subject is to be assigned both a row and a column label from the $r + c$ stacks of labels, f_1. of which are labeled row 1, f_2. of which are labeled row 2, and so forth. Let P denote the probability with which a specific table assembled at random will have these exact frequencies.

$$P = Q/R \text{ with}^3$$

---

[3] $\prod_{i=1}^{n} f_i! = f_1! f_2! \cdots f_n!$

$$Q = \prod_{i=1}^{r} f_{i.}! \prod_{j=1}^{c} f_{.j}! f_{..}! \quad \text{and } R = \prod_{i=1}^{r} \prod_{j=1}^{c} f_{ij}!$$

An obvious extension of Fisher's exact test is the Freeman and Halton [1951] test based on the proportion p of tables for which P is greater than or equal to P_o for the original table.

While the extension itself may be obvious, it's not as obvious this extension offers any protection against the alternatives of interest. Just because one table is less likely than another under the null hypothesis does not mean it is going to be more likely under the alternatives of interest to us. Consider the 1×3 contingency table $\boxed{f_1}\ \boxed{f_2}\ \boxed{f_3}$, which corresponds to the multinomial with probabilities $p_1 + p_2 + p_3 = 1$; the table whose entries are 1, 2, 3 argues more in favor of the null hypothesis $p_1 = p_2 = p_3$ than of the ordered alternative $p_1 > p_2 > p_3$.

The classic statistic for independence in a contingency table with r rows and c columns is

$$\chi^2 = \sum_{i=1}^{r} \sum_{j=1}^{c} (f_{ij} - Ef_{ij})^2 / Ef_{ij}$$

where Ef_{ij} is the number of observations in the ijth category one would expect on theoretical grounds.

With very large samples, this statistic has the chi-square distribution with $(r - 1)(c - 1)$ degrees of freedom. But in most practical applications, the chi-square distribution is only an approximation and is notoriously inexact for small and unevenly distributed samples.

The permutation statistic based on the proportion p_χ of tables for which χ^2 is greater than or equal to χ_o^2 for the original table provides an exact test and has all the advantages of the original chi-square. The distinction between the two approaches, as we observed in Chapter 6, is that with the original chi-square we look up the significance level in a table, while with the permutation statistic we derive the significance level from the permutation distribution. With large samples, the two approaches are equivalent, as the permutation distribution converges to the tabulated distribution (see chapter 14 of Bishop, Fienberg, and Holland [1975]).

This permutation test has one of the original chi-square test's disadvantages: While it offers global protection against a wide variety of alternatives, it offers no particular protection against any single one of them. The statistics p and p_χ treat row and column categories symmetrically, and no attempt is made to distinguish between cause and effect. To address this deficiency, Goodman and Kruskal [1954] introduce an asymmetric measure of association for nominal scale variables called τ, which measures the proportional reduction in error obtained when one variable, the "cause" or independent variable, is used to predict the other, the "effect" or dependent variable.

Assuming that the independent variable determines the row,

$$\tau = \frac{\sum_j f_{mj} - f_{m.}}{f_{..} - f_{m.}}$$

where $f_{mj} = \max_i f_{ij}$ and $f_{m.} = \max_i f_{i.}$.

$0 \leq \tau \leq 1$. $\tau = 0$ when the variables are independent; $\tau = 1$ when for each category of the independent variables all observations fall into exactly one category of the dependent. These points are illustrated in the following 2×3 tables:

3	6	9
6	12	18

$\tau = 0$

18	0	0
0	36	0

$\tau = 1$

3	6	9
12	18	6

$\tau = 0.166$

A permutation test of independence is based on the proportion of tables p_τ for which $\tau \geq \tau_o$.

Cochran's Q provides an alternative test for independence. Suppose we have I experimental subjects on each of whom we administer J tests. Let $y_{ij} = 1$ or 0 denote the outcome of the jth test on the ith patient, e.g., if the test is positive, $y_{ij} = 1$ and is 0 otherwise. Define $R_i = {}_j y_{ij}$; $C_j = {}_i y_{ij}$;

$$Q = \frac{\Sigma(C_j - C_.)^2}{R_. - \Sigma(R_i)^2}.$$

7.5. Ordered Statistical Tables

When data is measured on a continuous basis, such as 1.12, 1.13, 1.14, ties are a relatively infrequent occurrence. But when we ask someone to provide a self-rating on a discrete ordinal scale, 1 through 5, for example, ties are inevitable, the rule, rather than the exception, and the methods of this chapter may be more appropriate for analyzing such ordinal data than those of Chapter 3.

Which Test?

Use Fisher's exact test if:
    The data are in categories.
    The categories can't be ordered.
    There are exactly two rows and two columns.
Use the Freedman-Halton test or the chi-square if:
    There are more than two rows and at least two columns.
    You want to test whether the relative frequencies are the same in each row and in each column.
Use τ or Q if:
    You want to test whether the column frequencies depend on the row.

TABLE 7.5. Data Gathered by Graubard and Korn [1987]

Malformation	Maternal Alcohol Consumption (drinks/day)					Total
	0	< 1	1–2	3–5	≥ 6	
Absent	17,066	14,464	788	126	37	32,481
Present	48	38	5	1	1	93
	17,114	14,502	793	127	38	32,574

7.5.1. Ordered $2 \times c$ Tables

Our analysis of a $2 \times c$ ordered contingency table is straightforward and parallels the approach used in Section 3.4 for a k-sample comparison, once we have determined what value to assign each of the ordered categories. We illustrate this dilemma with data gathered by Graubard and Korn [1987] (see Table 7.5).

Recall that our test statistic is $\Sigma g[j] f_{1j}$ where g is any monotone increasing function. Among the leading choices for a scoring method g are

i) the category number: 1 for the first category, 2 for the second, and so on.

ii) the midrank scores.

iii) scores determined by the user, the choice we made in Section 3.4 when we analyzed the micronucleii data.

Consider the following 1×2 contingency table

	Alcohol Consumption	
Drinks/day	0	1–2
Frequency	3	5

The category or equidistant scores are 1 and 2. The ranks of the 8 observations are 1 through 3 and 4 through 8, so that the midrank score of those in the first category is 2 and in the second it's 6. Our user-chosen scores, corresponding to alcohol consumption, are 0 and 1.5.

Using Testimate^{TM} to analyze the full Gaubard-Korn data set, we obtain p-values that range from the insignificant, 0.29 for midrank scores, to marginally significant, 0.10 for the equidistant scores, to highly significant, 0.01 for our user-chosen scores. A user-chosen score based on the user's knowledge of underlying cause and effect is always recommended because it will be the most effective at distinguishing between hypothesis and alternative.

7.5.2. More than Two Rows and Two Columns

Two cases need to be considered. The first is when the columns but not the rows of the table may be ordered (the other variable being purely categorical), and the second is when both columns and rows can be ordered.

TABLE 7.6. Response to Chemotherapy

	None	Partial	Complete
CTX	2	0	0
CCNU	1	1	0
MTX	3	0	0
CTX + CCNU	2	2	0
CTX + CCNU + MTX	1	1	4

SINGLY ORDERED TABLES

Our approach parallels that of Section 8.1.2.2 in which we describe a two-way analysis of variance. Our test statistic is

$$F_2 = \Sigma(T_i - \overline{T})^2 \text{ where } T_i = \Sigma g_j f_{ij}.$$

As in the case of the $2 \times c$ table our problem is in deciding on the appropriate scores $\{g_j\}$.

Table 7.6 provides tumor regression data for 5 chemotherapy regimes. As partial response corresponds to approximately 2 years in remission (about 100 weeks) and complete response to an average of 3 years (150 weeks), we assign scores of 0, 100, and 150 to the ordered response categories.

To use StatXact-3 to analyze the tumor data, we first click on **TableData** in the main menu, click on **Settings**, then enter **Column Scores**. To execute the analysis, we select in turn **Statistics, Singly Ordered R x C Table, ANOVA with Arbitrary Scores**, and **Exact Test Method**. The results are shown here.

```
Datafile: C:\SX3WIN\EXAMPLES\TUMOR.CY3
ANOVA TEST [That the 5 rows are identically distributed]
Statistic based on the observed data :
  The Observed Statistic = 7.507
Asymptotic p-value: (based on Chi-square distribution with
    4 df)
  Pr { Statistic .GE. 7.507 } = 0.0747
Monte Carlo estimate of p-value :
  Pr { Statistic .GE. 7.507 } = 0.0444
  99.00% Confidence Interval = ( 0.0350, 0.0538 )

Elapsed Time i0:1:20.58 (3200 tables sampled with starting
    seed 16229)
```

Although our estimated significance level is less than 0.0444, a 99% confidence interval for this estimate, based on a sample of 3,200 possible rearrangements, does include values greater than 0.05. We can narrow this confidence interval by sampling additional rearrangements. With a Monte Carlo of 10,000, our Monte Carlo estimate of p-value is 0.0434 with a 99.00% confidence interval of (0.0382, 0.0486). Of course, the calculations take 3 times as long. When I first perform a permutation test, I use as few as 400 to 1,600 simulations. If the results are equivocal, as they were in this example, then and only then will I run 10,000 simulations.

DOUBLY ORDERED TABLES

In an $r \times c$ contingency table conditioned on fixed marginal totals, the outcome depends on the $(r - 1)(c - 1)$ odds ratios

$$\phi_{ij} = \frac{\pi_{ij}\pi_{i+1,j+1}}{\pi_{i,j+1}\pi_{i+1,j}}$$

where π_{ij} is the probability of an individual being classified in row i and column j.

In a 2×2 table, conditional probabilities depend on a single odds ratio and hence one- and two-tailed tests of association are easily defined. In an $r \times c$ table there are potentially, $n = 2(r - 1)(c - 1)$ sets of extreme values, 2 for each of the $(r - 1)(c - 1)$ odds ratios. Hence, an omnibus test for no association, e.g., χ^2, might have as many as 2^n tails.

Following Patefield [1982], we consider tests of the null hypothesis of no association between row and column categories H:$\phi_{ij} = 1$ for all i, j against the alternative of a positive trend K:$\phi_{ij} \geq 1$ for all i, j.

The principal test statistic considered by Patefield [1982], also known as the linear-by-linear association test, is

$$\lambda = \Sigma\Sigma f_{ij}r_ic_j$$

where $\{r_i\}$ and $\{c_j\}$ are user-chosen row and column scores.[4]

AN EXAMPLE

An ongoing fear of many parents is that something in the environment—asbestos or radon in the walls of their house, or toxic chemicals in their air and groundwater—will affect their offspring. Table 7.7 is extracted from data collected by Siemiatycki and McDonald [1972] on congenital neural-tube defects. Eyeballing the gradient along the diagonal of this table one might infer that births of ancephalic infants occur in clusters. The question arises as to which measures of distance and time we should use as weights $\{r_i\}$ and $\{c_j\}$.

Mantel [1967] reports striking differences between one analysis of epidemiologic data in which the coefficients are proportional to the differences in position and a second approach (which he recommends) to the same data in which the coefficients are proportional to the reciprocals of these differences. Using Mantel's approach, a pair of infants born 5 kilometers and 3 months apart contribute $\frac{1}{3} * \frac{1}{5} = \frac{1}{15}$ to the correlation. Summing up the contributions from all pairs and then repeating the summing process for a series of random rearrangements, Siemiatycki

---

[4]This statistic is actually just another form of Mantel's U, perhaps the most widely used of all multivariate statistics; See Section 9.3.

TABLE 7.7. Incidents of Pairs of Ancepahalic Infants

km Apart	Months apart		
	< 1	< 2	< 4
< 1	39	101	235
< 5	53	156	364
< 25	211	652	1,516

and McDonald conclude that the clustering of ancephalic infants is not statistically significant.

Enter the data into StatXact-3, and select **Statistics**, **Doubly Ordered** $R \times C$ **Table**, **Linear-by-Linear**, and **Exact** from the menus. Using the weights (2,.67,.33) and (2, .4, .08), we verify that a table this extreme would occur less than 1 in 5,000 times by chance.

7.6. Summary

In this chapter, you were introduced to the concept of a contingency table with fixed marginals and were shown that you could test against a wide variety of general and specific alternatives by examining the resampling distribution of the appropriate test statistic. Among the test statistics you considered were Fisher's exact, Freedman-Halton, chi-square, τ, Q, Pitman's correlation, and linear-by-linear association. These latter two statistics were to be used when you could take advantage of an ordering among the categories.

7.7. To Learn More

Excellent introductions to the analysis of contingency tables may be found in Agresti [1990, 1992], and in the StatXact-3 manual authored by Mehta and Patel (see Appendix 4). Major advances in analysis by resampling means have come about through the efforts of Gail and Mantel [1977], Mehta and Patel [1983], Mehta, Patel, and Senchaudhuri [1988], Baglivo, Oliver, and Pagano [1988, 1992], and Smith, Forester and McDonald [1996].

Berkson [1978], Basu [1979], Haber [1987] and Mielke and Berry [1992] examine Fisher's exact test. The power of the Freeman-Halton statistic in the $r \times 2$ case is studied by Krewski, Brennan and Bickis [1984]. Details of the calculation of the distribution of Cochran's Q under the assumption of independence are given in Patil [1975]. For a description of some other, alternative statistics for use in $r \times c$ contingency tables, see Nguyen [1985].

To study a 2×2 table in the presence of a third covariate, see Bross [1964] and Mehta, Patel, and Gray [1985].

7.8.   Exercises

1. 40,500 babies born in 1981 in the United States died before they were 28 days old. Of these babies, 30,000 were white and 10,500 were nonwhite. Comment on the hypothesis that black kids have a better chance to survive in North America than white kids.

2. Your friend rolls a die 120 times. Each time, before she rolls, you concentrate and visualize the die coming up a six. The die actually lands six a total of 28 times. Are you psychic?

3. A preliminary poll conducted well before the 1996 U.S. Presidential election yielded the following results:

	Males	Females
Dole	65	35
Clinton	55	42

Are these differences significant?

4. Will encouraging your child promote his or her intellectual development? A sample of 100 children and their mothers were observed, and the children's IQs tested at 6 and 12 years. Results were as follows:

	Mothers Encourage Schoolwork		
	Rarely	Sometimes	Always
IQ increased	8	15	27
IQ decreased	30	9	11

  a. Do you plan to perform a one-tailed or two-tailed test?
  b. What is the significance level of your test?

5. Does 1,2 dichloro-ethane induce tumors? Consider the following data evaluated by Gart et al. [1986]:

	Tumor	No Tumor
Treated	15	21
Control	2	35

6. Holmes and Williams [1954] studied tonsil size in children to verify a possible association with the virus S. pyrogenes. Do you feel there is an association? How many rows and columns in the following contingency table? Which, if any, of the variables is ordered?

Tonsil Size by Whether Carrier of S. Pyrogenes			
	Not Enlarged	Enlarged	Greatly Enlarged
Noncarrier	497	560	269
Carrier	19	29	24

7. Referring to Table 7.2, if Sandoz excluded cite 15 from the calculations, could they safely combine the data from the remaining cites?

8. Does dress make the wo_man?

	Time to First Promotion (months)		
	Long	Average	Short
Poorly dressed	12	8	4
Well dressed	18	25	20
Very well dressed	13	19	31

9. Referring to the literature of your own discipline, see if you can find a case where an $r \times 2$ table with at least one entry smaller than 7 gave rise to a borderline p-value using the traditional chi-square approximation. Reanalyze this table using resampling methods. Did the authors use a one-tailed or a two-tailed test? Was their choice appropriate?

10. Do Supreme Court appointments follow a Poisson distribution? Actual United States Supreme Court appointments grouped in 5-year periods from 1800 to 1990 appear below:

Period	Actual	Poisson	Period	Actual	Poisson
1800–04	2		1900–04	2	
1805–09	2		1905–09	2	
1810–14	2		1910–14	6	
1815–19	0		1915–19	2	
1820–24	1		1920–24	4	
1825–29	2		1925–29	1	
1830–34	1		1930–34	3	
1835–39	5		1935–39	4	
1840–44	1		1940–44	5	
1845–49	3		1945–49	4	
1850–54	2		1950–54	1	
1855–59	1		1955–59	4	
1860–64	5		1960–64	2	
1865–69	0		1965–69	3	
1870–74	4		1970–74	3	
1875–79	1		1975–79	1	
1880–84	4		1980–84	1	
1885–89	3		1985–89	2	
1890–94	4		1990–94	4	
1895–99	2		1995–99		

Recall from Section 4.2 that if an observation X has a Poisson distribution, such that we may expect an average of λ events per interval, the probability that $X = k$ in a given interval is $k^{-1} \exp[-\lambda k)$ for $k = 0, 1, 2, \ldots$. To test the hypothesis that Supreme Court appointments have the Poisson distribution, proceed as follows:

a. Compute the arithmetic average and use it to estimate λ.

b. Use the Poisson formula to fill in the remaining column of the table.

c. Use a statistics program to determine the probability of observing a $2 \times k$ contingency table that is as or more extreme.

d. Would you use the same test to establish whether Supreme Court appointments are made once every two years on the average?

11. According to the *Los Angeles Times*, a recent report in the *New England Journal of Medicine* states that a group of patients with a severe bacterial infection of their bloodstream who received a single intravenous dose of a genetically altered antibody had a 30% death rate compared with a 49% death rate for a group of untreated patients. How large a sample size would you require using Fisher's exact test to show that such a percentage difference was statistically significant?

Before you start your calculations, determine whether you should be using a one-tailed or a two-tailed test.

12. Do the 4 hospitals considered in Exercise 1.16 have the same billing practices?

13. Is treatment A superior? Consult Jonge [1983] to be sure. [Hints: Are the categories ordered? Is this a one- or two-tailed test?]

Outcome	A	B
Improvement		
Marked	7	3
Moderate	15	9
Slight	16	14
No change	13	21
Worse	1	5

CHAPTER 8

Experimental Design and Analysis

Failing to account for or balance extraneous factors can lead to major errors in interpretation. In this chapter, you'll learn to block or measure all factors that are under your control and to use random assignment to balance the effects of those you cannot. You'll learn to design experiments to investigate multiple factors simultaneously, thus obtaining the maximum amount of information while using the minimum number of samples.

8.1. Noise in the Data

In Chapter 2, we encountered a wo_man anxious to demonstrate her abilities as a tea taster. We argued in that chapter that by removing all extraneous sources of variation—e.g., the appearance of the cup, the temperature of the water, the expression on the experimenter's face—we could focus more narrowly on the factor to be tested.

Eliminating or reducing extraneous variation is the first of several preventive measures we use each time we design an experiment or survey. We strive to conduct our experiments in a biosphere with atmosphere and environment totally under our control. And when we can't—which is almost always the case—we record the values of the extraneous variables to use them either as *blocking* units or as *covariates*.

8.1.1. Blocking

Although the significance level of a permutation test may be distribution-free, its power (defined in Section 6.1.3) strongly depends on the underlying distribution.

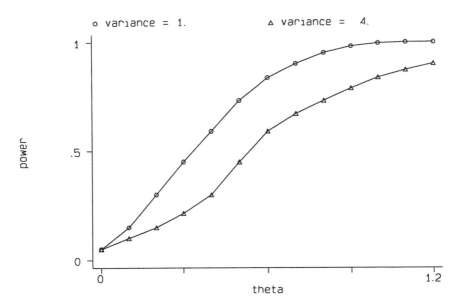

FIGURE 8.1. Power as a Function of the Population Variance.

In Figure 8.1, we see that the more variable our observations, the less the power of our tests and our ability to detect the alternative. One way to reduce the variance is to *block* the experiment, that is, to subdivide the population into more homogeneous subpopulations and to take separate independent samples from each.

Suppose you were designing a survey on the effect of income level on the respondent's attitude toward compulsory pregnancy. Obviously, the views of men and women differ markedly on this controversial topic. To reduce the variance and increase the power of your tests, *block* the experiment, interviewing and reporting on men and women separately. A physician would want to block by gender in a medical study, and probably by age and race as well. An agronomist would want to distinguish among clay soil, sand, and sandy-loam.

Whenever a population can be subdivided into distinguishable subpopu-lations, you can reduce the variance of your observations and increase the power of your statistical tests by blocking or stratifying your sample.

Suppose we have agreed to divide our sample into two blocks—one for men and one for women. If this is an experiment, rather than a survey, we would then assign subjects to treatments separately and independently within each block. In a study that involves 2 treatments and 10 experimental subjects, 4 men and 6 women, we would first assign the men to treatment and then the women. We could assign the men in any of $\binom{4}{2} = 6$ ways and the women in any of $\binom{6}{3} = 20$ ways, for a total of $6 \times 20 = 120$ possible random assignments.

When we come to analyze the results of our experiment, we use the permutation approach to ensure that we analyze the way the experiment was designed. Our test statistic is a natural extension of that used for the two-sample comparison in that

---

How Many Controls?

Every experiment involves at least two groups of subjects—those that took the drug versus those who took the placebo (the control group), or those that took the new drug versus those that took the old.

For the self-assured, the smug, and the person with something to sell, controls may be an unnecessary luxury. But the day I started at the Upjohn Company, my boss John R. Schultz recommended I use twice as many subjects in each control group as the number devoted to experimental treatment and that I use two types of controls, positive and negative.

For a study on Motrin, the positive control was aspirin (the best treatment at the time) and the negative or neutral control was a harmless filler used to give aspirin tablets their compact shape. Why did we use so many subjects in each control group? "Life is full of surprises," said John. "You leave work one day whistling, the next day you're back with a head cold. Most of the time these negative effects have nothing to do with the treatment. By using many control subjects, you ensure the normal wear and tear of ordinary life will be detected and accounted for and won't be falsely associated with the treatment you're trying to investigate."

Further proof of John's wisdom came several years later with the controversy over silicon implants. Women who had the implants suffered from a wide variety of complaints. Dow Corning, the manufacturer, didn't have the data on control groups to show that such complaints might be pure coincidence and was forced to pay millions of dollars in compensation.

---

we simply sum the separate sums computed for each block

$$S = \sum_{b=1}^{B} \sum_{j=m_b+1}^{(n_b+m_b)} x_{bj} \tag{8.1}$$

where B is the number of blocks (2 in the current example) and the inner sum extends over the n_b-treated observations $\{x_{bj}\}$ within each block.

We compute the test statistic for the original data. Then we rearrange the observations at random independently within each block, subject to the restriction that the number of observations within each treatment category—in this example, the pair $\{n_b, m_b\}$—remains constant.

We compute S for each of the 120 possible rearrangements. If the value of S for the original data is among the 120α largest values, then we reject the null hypothesis at the α significance level; otherwise we accept it.

The resulting permutation test is exact and most powerful against normal alternatives even if the observations on men and women have different distributions [Lehmann, 1986]. As we saw in Section 6.2, all that is required is that the subsets of errors be exchangeable.

---

[1] Recall that $\sum_{k=1}^{n} X_k = X_1 + X_2 + \cdots X_n$ where X_i is the ith observation.

Analyzing A Blocked Experiment

Compute the test statistic for the original data, for example,

$$S_o = \sum_{b=1}^{B} \sum_{j} x_{bj}$$

where the inner sum extends over the first sample in the block.

Repeat the following 400 times:
  For each block:
    Rearrange the data in the block.
    Compute S.
    Record the number of times S exceeds S_0.

If the proportion of rearrangements in which S exceeds S_0 is small, conclude that the difference between the samples is statistically significant.

The design need not be balanced. The test statistic S (equation 8.1) is a sum of several independent sums. Unequal sample sizes resulting from missing data or an inability to complete one or more portions of the experiment will affect the analysis only in the relative weights assigned to each subgrouping.[2]

Blocking is applicable to any number of subgroups; the extreme case in which every pair of observations forms a distinct subgroup is *matched pairs*, which we studied in Section 3.9.

8.1.2. Measuring Factors We Can't Control

Many of the factors in an experiment will be beyond our control. Rainfall in an agricultural experiment is one example. If we are studying the effects of advertising on the sale of beer, a sudden heat wave or an unexpected drop in temperature will markedly effect our results. But we can measure rainfall and temperature and use our observations to either block or correct our results.

We could categorize rainfall as light, normal, or heavy and, as in Section 2.2, we can incorporate rainfall in some kind of model. For example, if X_i denotes the yield of one of our test acres, we might want to consider the corrected yield $X_i' = X_i - a - bR_i$ where R_i is the quantity of rain that fell on that acre and a and b are constants in our model. We would then apply our permutation procedures to the corrected values.

---

[2] *Warning*: This remark applies only if the data is missing at random. If treatment-related withdrawals are a problem in one of your studies, see Entsuah [1990] for the details of a resampling procedure.

Experimental Design

List factors you feel may influence the outcome of your experiment.
Block factors that are under your control. You may want to use some of these factors to restrict the scope of your experiment, e.g., eliminate all individuals under 18 and over 60.
Measure the remaining factors.
Randomly assign units to treatment within each block.

8.1.3. Randomization

Imagine you are ready for an experiment to find a new and better fertilizer for growing tomatoes. You've set up the experiment indoors under the lights so you can keep total control over duration of daylight, moisture, and temperature. Because you are hunting for a general-purpose fertilizer, you've put out three separate sets of trays containing sandy loam, a sand-clay mix, and a clay soil. Now, it just remains to put the plants in the soil. Be careful. Tomato plants are not all alike, and there are big differences among seedlings. Studies have shown there is a natural tendency to plant the tall seedlings first and save the runts for last. This is not a good idea if all the runts end up in one tray and all the big sturdy plants in another.

An alternate plan would be to deal the seedlings out like cards, the first for Tray 1, the second for Tray 2, and so on down the line. Alas, this method would still seem to put the better plants into the first tray.

When we analyze this experiment using permutation methods, we assume that each assignment of labels to treatments is equally likely. Only if we assign the plants to the different trays completely at random will this assumption be fulfilled. If the plants assigned to Tray 1 prove in the end to be larger on the average than those in Tray 2, this should be strictly a matter of chance, not the result of a specific bias on our part.

Randomly assign subjects to treatment whenever you don't or can't control all the factors in an experiment.

Our tomato study might use one of the following randomization schemes:

1. Choose a random rearrangement of the integers, $1 \ldots m$ corresponding to the m trays. Place the first m plants in the trays in the order specified. Choose a second random rearrangement, then a third, and so on, until all the plants have been placed in the trays.
2. Throw an m-sided die and place the first plant in the indicated tray. Continue to throw the die until all the plants have been placed. If the die indicates that an already full tray has been selected, just ignore it and throw again. (See also Exercise 4.)

8.2.   Balanced Designs

What distinguishes the complex experimental design from the simple one-sample, two-sample, and k-sample experiments we considered in Chapter 3 is the presence of multiple control factors.

For example, we may want to assess the simultaneous effects on crop yield of hours of sunlight and rainfall. We determine to observe the crop yield X_{ijm} for I different levels of sunlight, $i = 1, \ldots I$, and J different levels of rainfall, $j = 1, \ldots J$, and to make M observations at each factor combination, $m = 1, \ldots M$. We adopt as our model relating the dependent variable crop yield (the effect) to the independent variables of sunlight and rainfall (the causes)

$$X_{ijm} = \mu + s_i + r_j + (sr)_{ij} + \epsilon_{ijm}.$$

In this model, terms with a single subscript, like s_i, the effect of sunlight, are called *main effects*. Terms with multiple subscripts, like $(sr)_{ij}$, the residual and nonadditive effect of sunlight and rainfall, are called *interactions*. The *residuals* $\{\epsilon_{ijm}\}$ represent that portion of crop yield that cannot be explained by the independent variables alone. To ensure that the residuals are exchangeable so that permutation methods can be applied, the experimental units must be assigned at random to treatment.

Suppose we also want to compare the effects of several different fertilizers for different combinations of sunlight and rainfall. We would observe the crop yield X_{ijkm} for I different levels of sunlight, J different levels of rainfall, and K different levels of fertilizer, and make M observations at each factor combination. Our model would then be

$$X_{ijkm} = \mu + s_i + r_j + f_k + (sr)_{ij} + (sf)_{ik} + (rf)_{jk} + (srf)_{ijk} + \epsilon_{ijkm}.$$

In this model we have three main effects, s_i, r_j, and f_k, three two-way interactions, $(sr)_{ij}, (sf)_{ik}$, and $(rf)_{jk}$, a single three-way interaction, $(srf)_{ijk}$, and the error term ϵ_{ijkm}.

Including the additive constant μ in the model allows us to define all main effects and interactions so they sum to zero, $\Sigma s_i = 0$; $\Sigma r_j = 0$; $\Sigma f_k = 0$; $\Sigma\Sigma(sr)_{ij} = 0$; $\Sigma\Sigma(sf)_{ik} = 0$; $\Sigma\Sigma(rf)_{jk} = 0$; and $\Sigma\Sigma\Sigma(srf)_{ijlk} = 0$. As a result, the null hypothesis of no effect of sunlight on crop yield is equivalent to the statement that each of the main effects $s_i = 0$ for $i = 1, \ldots I$. Under the alternative, the different terms s_i represent deviations from a zero average, with the interaction term $(sr)_{ij}$ representing the deviation from the sum $s_i + r_j$.

When we have multiple factors, we must also have multiple test statistics. In the preceding example, we require three separate tests and test statistics for the three main effects s_i, r_j, and f_k, plus four other statistical tests for the three two-way and the one three-way interactions. Will we be able to find statistics that measure a single intended effect without *confounding* it with a second unrelated effect? Will the several test statistics be independent of one another?

The answer is *yes* to both questions only if the design is *balanced*, that is, if there are equal numbers of observations in each subcategory, and if the test statistics are independent of one another. In an unbalanced design, main effects will be confounded with interactions so the two cannot be tested separately, a topic we'll return to with a bootstrap solution in Section 8.5.

8.2.1. Main Effects

In a k-way analysis with equal sample sizes M in each category, we can assess the main effects using essentially the same statistics we would use for randomized blocks. Take sunlight in the preceding example. If we have only two levels of sunlight, then, referring to Equation 8.1, our test statistic for the effect of sunlight is

$$S = \sum_j \sum_k \sum_m X_{1jkm}, \tag{8.2}$$

the sum of all observations at the first level of sunlight.

If we have more than two levels of sunlight, our test statistic is

$$F_2 = \sum_i \sum_j \sum_k (\overline{X}_{ijk.} - \overline{X}_{.jk.})^2 \tag{8.3}$$

or

$$F_1 = \sum_i \sum_j \sum_k |\overline{X}_{ijk.} - \overline{X}_{.jk.}|^3 \tag{8.4}$$

Whether we use F_1 or F_2 will depend on our loss function (see Section 6.1.4) and the relative weight we assign to far versus near alternatives. F_2 would be our choice, for example, when observations are expensive or difficult to obtain and we want to increase the probability of detecting distant alternatives even if it means decreasing the probability of detecting near ones. We may also want to substitute medians for means or replace the original observations by their ranks or some other transformation (see Section 3.8).

The statistics F_2 and F_1 offer protection against a variety of shift alternatives, including

$$K_1 : s_1 = s_2 > s_3 = \cdots$$

---

[3] The dot (.) used as a subscript indicates we have summed over the corresponding subscript; thus

$$X_{ijk.} = \sum_{m=1}^{M} X_{ijkm}$$

The bar over the letter indicates we have taken the average; thus

$$\overline{X}_{ijk.} = \frac{1}{M} \sum_{m=1}^{M} X_{ijkm}.$$

$$K_2 : s_1 > s_2 > s_3 = \cdots$$
$$K_3 : s_1 < s_2 > s_3 = \cdots$$

As a result, they may not provide a most powerful test against any alternative. If we believe the effect to be monotone increasing, then, in line with the thinking detailed in Section 3.4, we would use the Pitman correlation statistic:

$$R = \sum_i \sum_j \sum_k f(i)\overline{X}_{ijk}. \qquad (8.5)$$

To obtain the permutation distributions of the test statistics S, F_2, F_1, and R for the effect of sunlight, we permute the observations independently in each of the JK blocks determined by a specific combination of rainfall and fertilizer. Exchanging observations within a category corresponding to a specific level of sunlight i leaves the statistics S, F_2, F_1, and R unchanged. We can concentrate on exchanges between categories and the total number of rearrangements is $JK\binom{IM}{M \cdots M}$.

We compute the test statistic (S, F_1, or R) for each rearrangement, rejecting the hypothesis that sunlight has no effect on crop yield only if the value of S (or F_1 or R) that we obtain using the original arrangement of the observations lies among the α most extreme of these values.

Of the two F-statistics, F_1 is to be preferred to F_2. F_1 is as or more powerful for detecting location shifts and more powerful for detecting concentration changes [Mielke, 1986].

A final alternative to the statistics S, F_1, and F_2 is the standard F-ratio statistic

$$F = \frac{MJK \sum_i (\overline{X}_{i..} - \overline{X}_{...})^2}{(I - 1)\widehat{\sigma}^2} \qquad (8.6)$$

where $\widehat{\sigma}^2$ is our estimate of the variance of the errors ϵ_{ijkm}. But if we use F, we are forced to consider exchanges between as well as within blocks, thus negating the advantages of blocking.

ANALYZING A ONE-WAY TABLE

Our first example concerns a straightforward comparison of the effects of three brands of fertilizer on crop yield as recorded in Table 8.1a.

We'll use F_1 as our test statistic, which in this example reduces to

$$|\overline{X}_{1.} - \overline{X}_{..}| + |\overline{X}_{2.} - \overline{X}_{..}| + |\overline{X}_{3.} - \overline{X}_{..}|.$$

TABLE 8.1. Effects of Fertilizer on Crop Yield

FastGro	NewGro	WunderGro
27	38	75
30	12	76
55	72	54
72		
18		

TABLE 8.2. Effects of Fertilizer on Crop Yield; Deviations from Grand Mean

FastGro	NewGro	WunderGro
−17	−6	+31
−14	−32	+32
+11	+28	+10
+28		
−26		

TABLE 8.3. Effects of Fertilizer on Crop Yield; Rearranged Deviations

FastGro	NewGro	WunderGro
−17	−6	+31
−14	−32	+32
+11	+28	−26
+28		
+10		

$\overline{X}_{..} = 44$; to simplify our calculations, we'll first subtract 44 from each observation. F_1 reduces a second time to the sum of the absolute values of the means for each type of fertilizer, that is, $F_1 = |\overline{X}_{1.}| + |\overline{X}_{2.}| + |\overline{X}_{3.}|$, or $18 + 10 + 73 = 101$ for Table 8.1b.

In Table 8.1c, we have rearranged the observations, preserving the number of observations in each category and the new value of F_1 is $18 + 10 + 37 = 65$.

Continuing in this fashion until we have considered all possible rearrangements of the observations among the categories, $\binom{11}{5\ 3}$, we see a value as large as 101 occurs less than 1% of the time, and we conclude that there is a statistically significant difference among the fertilizers.[4] A C++ program to do these calculations is included in Appendix 2.

ANALYZING A TWO-WAY TABLE

In this second example, we apply the permutation method to determine the main effects of sunlight and fertilizer on crop yield using the data from the two-factor experiment depicted in Table 8.2a. There are only two levels of sunlight in this experiment, so we use S (Equation 8.1) to test for the main effect.

For the original observations, $S = 23 + 55 + 75 = 153$. One possible rearrangement is shown in Table 8.2b, in which we have interchanged the two observations marked with an asterisk, the 5 and 6. The new value of S is 154.

As can be seen by a continuing series of straightforward hand calculations, the test statistic S for the main effect of sunlight is as small as or smaller than it is for the original observations in only 8 of the $\binom{6}{3}^3 = 8,000$ possible rearrangements. For example, it is smaller when we swap the 9 of the Hi-Lo group for the 10 of the

---

[4]Recall from Section 2.2.2.1 that $\binom{11}{5\ 3}$ means $\frac{11!}{5!3!(11-8)!}$.

Analyzing a One-Way Table

H: mean/median the same for all treatments.
K: means/medians are different for at least one treatment.

Assumptions:

1. Observations are exchangeable if the hypothesis is true.
2. Choose a test statistic:
   Treatments cannot be ordered.
      Even small differences in means are important — use F_1.
      Only large differences in means are important — use F_2.
   Treatments can be ordered — use R.
3. Compute the test statistic for the original data.
4. Repeat 400 times:
      Rearrange the data.
      Compute test statistic.
      Record the number of times the statistic exceeds the original value.

TABLE 8.4. Effect of Sunlight and Fertilizer on Crop Yield

		Lo	Med	High
S			Fertilizer	
U	Lo	5	15	21
N		10	22	29
L		8	18	25
I				
G	Hi	6	25	55
H		9	32	60
T		12	40	48

TABLE 8.5. Effect of Sunlight and Fertilizer; Data Rearranged

	Lo	Med	High
Lo	6*	15	21
	10#	22	29
	8	18	25
Hi	5*	25	55
	9#	32	60
	12	40	48

Lo-Lo group (the two observations marked with the pound sign, #). As a result, we conclude that the effect of sunlight is statistically significant.

The computations for the main effect of fertilizer are more complicated—we must examine $\left(\begin{smallmatrix} 9 \\ 3 \ 3 \end{smallmatrix}\right)^2$ rearrangements, and compute the statistic F_1 for each. We use

F_1 rather than R because of the possibility that too much fertilizer—the "High" level—might actually suppress growth. Only a computer can do this many calculations quickly and correctly, so we adapted our C++ program from Appendix 2 to make them. The estimated significance level is .001 and we conclude that this main effect, too, is statistically significant.

In this last example, each category held the same number of experimental subjects. If the numbers of observations were unequal, our main effect would have been confounded with one or more of the interactions (see Section 8.5). In contrast to the simpler designs we studied in Chapter 3, missing data will affect our analysis.

8.2.2. Testing for Interactions

In the preceding analysis of main effects, we assumed the effect of sunlight was the same regardless of the levels of the other factors. To test the validity of this assumption, we first eliminate row and column effects by subtracting the row and column means from the original observations. That is, we set

$$X'_{ijk} = X_{ijk} - \overline{X}_{i..} - \overline{X}_{.j.} + \overline{X}_{...};$$

where by adding the grand mean, $\overline{X}_{...}$, we ensure the overall sum will be zero. In the example of the effect of sunlight and fertilizer on crop yield, we are left with the residuals shown in Table 8.6.

The pattern of plus and minus signs in this table of residuals suggests that fertilizer and sunlight affect crop yield in a superadditive fashion. Note the minus signs associated with the mismatched combinations of a high level of sunlight and a low level of fertilizer and a low level of sunlight with a high level of fertilizer. To encapsulate our intuition in numeric form, we sum the deviates from the mean within each cell, square the sum, and then sum the squares to form the test statistic

$$I = \sum_i \sum_j \left(\sum_k X'_{ijk} \right)^2. \tag{8.7}$$

TABLE 8.6. Effect of Sunlight and Fertilizer on Crop Yield; Testing for Nonadditive Interaction

S			Fertilizer	
		Lo	Med	High
U	Lo	4.1	−2.1	−11.2
N		9.1	4.1	−3.2
L		7.1	0.1	−7.2
I				
G	Hi	−9.8	−7.7	7.8
H		−7.8	−0.7	12.8
T		−3.8	7.2	0.8

We compute this test statistic for each rerandomization of the 18 deviates into 6 subsamples. In most, the values of the test statistic are close to zero because the entries in each cell cancel. The value of the test statistic for our original data, $I = 2127.8$, stands out as an exceptional value and we conclude that there is a significant interaction between sunlight and fertilizer ($\alpha < .003$).

As the deviates are weakly correlated, hence, not exchangeable, this significance level is only an approximation. Erring on the conservative side, we conclude the two variables do not act independently, and, in reporting our results, always refer to the combined effect of sunlight and fertilizer.

A C++ program to test interactions is given in Appendix 2.

8.3. Designing an Experiment

All the preceding results are based on the assumption that the assignment of treatments to plots (or subjects) is made at random. While it might be convenient to fertilize our plots as shown in Table 8.4a, the result could be a systematic bias, particularly if, for example, there is a gradient in dissolved minerals from east to west across the field. The layout adopted in Table 8.4b, obtained with the aid of a computerized random number generator, reduces but does not eliminate the effects of this hypothetical gradient. Because this layout was selected at random, the exchangability of the error terms and, hence, the exactness of the corresponding permutation test is assured. Unfortunately, the layout of Table 8.4a with its built-in bias can also result from a random assignment; its selection is neither more nor less probable than any of the other $\binom{9}{3\ 3}$ possibilities.

What can we do to avoid such an undesirable event? In the layout of Table 8.4c, known as a *Latin square*, each fertilizer level occurs once and once only in each row and in each column; if there is a systematic gradient of minerals in the soil, then this layout ensures the gradient will have almost equal impact on each of the three treatment levels. It will have an almost equal impact even if the gradient extends from northeast to southwest rather than from east to west or north to south. I use the phrase "almost equal" because a gradient effect may still persist. The design and analysis of Latin squares is described in the next section.

8.3.1. Latin Square

The Latin square considered in Section 8.3 is one of the simplest examples of an experimental design in which the statistician takes advantage of some aspect of the model to reduce the overall sample size.

A Latin square is a three-factor experiment in which each combination of factors occurs once and once only. We can use a Latin Square as in Table 8.4 to assess the effects of soil composition on crop yield (see Table 8.5).

Analyzing a Two-Way Table

H_1: Treatment effects are the same for all levels of the other factor.
K_1: Treatment effect depends on the level of the other factor.

Assumptions:

1. Residuals are exchangeable if the hypothesis is true.

Procedure:

1. Compute cell means $\{\overline{X}_{ij.}\}$, row means $\{\overline{X}_{i..}\}$, column means $\{\overline{X}_{.j.}\}$, and grand mean $\overline{X}_{...}$.
2. Calculate residuals.
3. Compute the interaction statistic I (Equation 8.7).
4. Obtain the permutation diistribution of this statistic.
5. If I_o is an extreme value, stop. Effects must be estimated separately for each treatment combination.
6. If I_o is not an extreme value, you may proceed to test for main effects.

H_2: Treatment has no effect.
K_2: Outcome depends on the treatment.

Assumptions:

1. Observations are exchangeable if the hypothesis is true.

Procedure:

For *each* of the treatments:
Choose a test statistic.
Treatment levels can be ordered.
Even small differences in means are important — use F_1.
Only large differences in means are important — use F_2.
Treatment levels can be ordered — use R.
Compute the statistic for the original observations.
Obtain permutation distribution of the statistic.
Consider only rearrangements that result in exchanging values among treatment levels.
Draw a conclusion.

TABLE 8.7. Systematic Assignment of Fertilizer Levels to Plots

Hi	Med	Lo
Hi	Med	Lo
Hi	Med	Lo

TABLE 8.8. Random Assignment of Fertilizer Levels to Plots

Hi	Med	Lo
Lo	Med	Lo
Hi	Hi	Lo

TABLE 8.9. Latin Square Assignment of Fertilizer Levels to Plots

Hi	Med	Lo
Lo	Hi	Med
Med	Lo	Hi

In this diagram, Factor 1, gypsum concentration, say, is increasing from left to right, Factor 2 is increasing from top to bottom (or from north to south), and the third factor, its varying levels denoted by the capital letters A, B, and C, occurs in combination with the other two in such a way that each combination of factors—row, column, and treatment—occurs once and once only.

Because of this latter restriction, there are only 12 different ways in which we can assign the varying factor levels to form a 3×3 Latin square. Among the other 11 designs are

	1	2	3
1	A	C	B
2	B	A	C
3	C	B	A

and

TABLE 8.10. A Latin Square

F		**Factor 1**		
a		1	2	3
c	1	A	B	C
t	2	B	C	A
o	3	C	A	B
r				

1	C	B	A
2	B	A	C
3	A	C	B

We assume we begin our experiment by selecting one of these 12 designs at random and planting our seeds in accordance with the indicated conditions.

Because there is only a single replication of each factor combination in a Latin square, we cannot estimate the interactions. The Latin square is appropriate only if we feel confident in assuming the effects of the various factors are completely additive, that is, the interaction terms are zero.

Our model for the Latin square is

$$X_{ijk} = s_i + r_j + f_k + \epsilon_{ijk}$$

where, as always in a permutation analysis, we assume that the errors $\{\epsilon_{ijk}\}$ are exchangeable. Our null hypothesis is $H : s_1 = s_2 = s_3$. If we assume an ordered alternative, $K : s_1 > s_2 > s_3$, our test statistic for the main effect is similar to the correlation statistic (Equation 8.5):

$$R = \sum_{i=1}^{3}(i - 1)(\overline{X}_{i..} - \overline{X}_{...})$$

or, equivalently, after eliminating the grand mean $\overline{X}_{...}$, which is invariant under permutations, and multiplying by n, $R' = -X_{A..} + X_{C..}$.

We evaluate this test statistic for both the observed design and each of the 12 possible Latin square designs that might have been used in this particular experiment. We reject the hypothesis of no treatment effect only if the test statistic for the original observations is an extreme value.

For example, suppose we used Design 1 and observed

21	28	17
14	27	19
13	18	23

Then $X_{A..} = 58$, $X_{C..} = 57$, and our test statistic $R' = -1$. Had we used Design 2, then $X_{A..} = 71$, $X_{C..} = 65$, and our test statistic $R' = -7$, while with Design 3, $X_{A..} = 57$, $X_{C..} = 58$, and our test statistic is $+1$.

We see from the permutation distribution obtained in this manner that -1, the value of our test statistic for the design actually used in the experiment, is an average value, not an extreme one. We accept the null hypothesis and conclude that increasing the treatment level from A to B to C does not significantly increase the yield.

8.4. Another Worked-Through Example

Consider the following example suggested by Tony DuSoir, the author of SC, which is taken from Hettmansperger [1984]. Survival times are measured for each

---

Designing and Analyzing a Latin Square

H: mean/median the same for all levels of each treatment.
K: means/medians are different for at least one level.

Assumptions:

1. Observations are exchangeable if the hypothesis is true.
2. Treatment effects are additive (not synergistic or antagonistic).

Procedure:

1. List all possible Latin Squares for the given number of treatment levels.
2. Assign one at random and use it to perform the experiment.
3. Choose a test statistic (R or F_1).
4. Compute the statistic for the design you used.
5. Compute the test statistic for all the other possible designs.
6. Determine whether the original value of the test statistic is extreme.

---

of 3 different poisons and 4 different treatments. The experiment is replicated 4 different times and the results are displayed in the following table.

		Treatment			
		1	2	3	4
P	1	31, 45, 46, 43	82, 110, 88, 72	43, 45, 63, 76	45, 71, 66, 62
I	2	36, 29, 40, 23	92, 61, 49, 124	44, 35, 31, 40	56, 102, 71, 38
N	3	22, 21, 18, 23	30, 37, 38, 29	23, 25, 24, 22	30, 36, 31, 33

The mean survival times of the four treatment subgroups are: 31.4, 67.6, 39.2, and 53.4 days, respectively. The mean survival times of the three poison subgroups are: 61.8, 54.4, and 27.6 days, respectively. The grand mean is 47.9.

This experiment is balanced, and our first step is to check for interactions; we subtract the treatment and poison means and add the grand mean to each observation to obtain the following table of residuals.

		Treatment			
		1	2	3	4
P	1	−14, 0, 1, 2	1, 30, 8, −9	−10, −8, 10, 23	−22, 5, −1, −5
I	2	−2, −9, 2, −15	18, −13, −25, 46	−2, −11, −15, −6	−4, 42, 11, −22
N	3	11, 10, 7, 12	−17, −10, −9, −18	4, 6, 5, 3	−3, 3, −2, 0

The statistic $I = 3.5$; a value this extreme occurs in less than 3% of the rearrangements, suggesting that the effects of treatment and poison are not additive.[5]

We cannot test to see if the main effects are significant because they are confounded with the interactions. We can test to see whether there are significant differences among treatments when our attention is restricted to one of the poisons, as in the following table.

	Treatment			
	1	**2**	**3**	**4**
1	31, 45, 46, 43	82, 110, 88, 72	43, 45, 63, 76	45, 71, 66, 62

The treatment means are 41.25, 88.0, 56.75, and 61.0; the grand mean is 61.75. The statistic F_1, the sum of the absolute deviations about the grand mean, is 52.5. The probability a value this large will occur purely by chance is less than 1%. We conclude there are significant differences among treatments when poison 1 is used.

8.5. Determining Sample Size

Power is an increasing function of sample size, as we saw in Figures 6.2 and 6.3, so we should take as large a sample as we can afford, providing the gain in power from each new observation is worth the expense of gathering it.

For one- and two-sample comparisons of means and proportions, a number of commercially available statistics packages can help us with our determination. We've included a sample calculation with Stata in a sidebar. Note that to run Stata or similar programs, we need to specify in advance the alternative of interest and the desired significance level.

To determine sample size for more complicated experimental designs, we need to run a computer simulation. Either we specify the underlying distribution on theoretical grounds—as a normal or mixture of normals, a γ, a Weibull or some other well-tabulated function—or we use the empirical distribution obtained in some earlier experiment.

We also need to specify the significance level, e.g., 5%, and the alternative of interest, e.g., values in the second sample average 2 points larger than in the first.

To simulate the sampling process for a known distribution like the normal, we first choose a uniformly distributed random number between 0 and 1, the same range of values taken by the distribution function, then we look this number up as an entry in a table of the inverse of the normal distribution.

Programming in Stata, for example, we write invnorm(uniform()) and repeat this command for each element in the untreated sample. If our alternative is that the population comes from a population with mean 5 and standard deviation 2,

---

[5] By contrast, the standard parametric (ANOVA) approach yields a p-value greater than 10%.

Using Stata to Determine Sample Size

Study of effect of oral contraceptives (OC) on blood pressure of women ages 35-39.

OC Users 133 ± 15
NonUsers 127 ± 18

. sampsi 133 127, alpha(0.05) power(0.8) sd1(15) sd2(18)

Estimated sample size for two-sample comparison of means

Test Ho: m1 = m2, where m1 is the mean in population 1 and m2 is the mean in population 2

Assumptions:
  alpha = 0.0500 (two-sided)
  power = 0.8000
  m1 = 133
  m2 = 127
  sd1 = 15
  sd2 = 18
  n2/n1 = 1.00

Estimated required sample sizes:
  n1 = 120
  n2 = 120

---

we would write $5 + 2*\text{invnorm}(\text{uniform}())$. We repeat this command for each observation in the treated sample.

If we aren't sure about the underlying distribution, we draw a bootstrap sample with replacement from the empirical distribution.

We compute the test statistic for the sample and note whether we accept or reject.

We repeat the entire process 50 to 400 times (50 times when just trying to get a rough idea of the correct sample size, 400 times when closing in on the final value). The number of rejections divided by the number of simulations provides us with an estimate of the power for the specific experimental design and our initial sample sizes. If the power is still too low, we increase the sample sizes and repeat the preceding simulation process.

8.6.   Unbalanced Designs

Imbalance in the design will result in the *confounding* of main effects with interactions. Consider the following two-factor model for crop yield:

$$X_{ijk} = \mu + s_i + r_j + (sr)_{ij} + \epsilon_{ijk}.$$

Now, suppose the observations in a two-factor experimental design are distributed as in the following diagram:

$$\frac{\text{Mean } 0| \text{ Mean } 2}{\text{Mean } 2| \text{ Mean } 0}$$

There are no main effects in this example—both row means and both column means have the same expectation—but there is a clear interaction represented by the two nonzero off-diagonal elements.

If the design is balanced, with equal numbers per cell, the lack of significant main effects and the presence of a significant interaction should and will be confirmed by our analysis. But suppose the design is not in balance, that for every 10 observations in the first column, we have only 1 observation in the second. Because of this imbalance, when we use the statistic S (Equation 8.1), we will uncover a false "row" effect that is actually due to the interaction between rows and columns. The main effect is said to be *confounded* with the interaction.

If a design is unbalanced as in the preceding example, we cannot test for a "pure" main effect or a "pure" interaction. But we may be able to test for the combination of a main effect with an interaction by using the statistic $(S, F_1, \text{ or } R)$ that we would use to test for the main effect alone. This combined effect will not be confounded with the main effects of other unrelated factors.

8.6.1.   Bootstrap to the Rescue

The bootstrap can help us derive confidence limits even with an unbalanced design. Here is an example taken from an experiment by Plackett and Hewlett [1963] in which milkweed bugs were exposed to various levels of two insecticides. At issue is whether the two drugs act independently.

Although death due to a variety of spontaneous and background causes could be anticipated, no attempt was made to actually measure this background—the cell corresponding to a zero dose of each drug is empty. The resultant design

TABLE 8.11. Deaths of Milkweed Bugs Exposed to Various Levels of 2 Insecticides; 48 Bugs per Group.

Dose B	Dose A 0	0.05	0.07
0		9	22
0.2	5	27	27

is unbalanced. Still, following the lead of Wåhrenlot [1980], we can bootstrap a solution.

An underlying biological assumption is that the dose threshold above which a given insecticide is toxic varies from insect to insect. Suppose we form a pair of bootstrap samples. The first sample we construct in two stages: First, we draw an observation at random from the sample of 48 milkweed bugs treated with 0.05 unit of the first insecticide alone. If by chance we select one of the 39 survivors, then we draw from the sample of 48 bugs treated with 0.2 unit of the second insecticide alone. Otherwise, we record a "death."

Of course, we don't actually perform the drawing but simulate it through the use of a random number generator. If this number is greater than 9/48, the insect lives to be treated a second time: otherwise it dies.

The second bootstrap sample we select with replacement from the 27 killed and 21 survivors in the sample treated with both insecticides. We repeat the process 50 to 200 times, each time comparing the number of survivors in the two bootstrap samples. If the two insecticides act independently, the numbers should be comparable. See, also, Romano [1988].

8.6.2.   Missing Combinations

If an entire factor-combination is missing, we may not be able to estimate or test any of the effects. One very concrete example is an unbalanced design I encountered in the 1970s when I worked with Makinodan et al. [1976] to study the effects of age on the mediation of the immune response. They measured the anti-SBRC response of spleen cells derived from C57BL mice of various ages. In one set of trials, the cells were derived entirely from the spleens of young mice: in a second, they came from the spleens of old mice: and in a third they came from mixtures of the two.

Let X_{ijk} denote the response of the kth sample taken from a population of type i, j where $i = 0 = j$ denotes controls; $i = 1, j = 0$ denotes cells from young animals only; $i = 0, j = 1$ denotes cells from old animals only; and $i = 1 = j$ denotes a mixture of cells from old and young animals. We assume for lymphocytes taken from the spleens of young animals that

$$X_{10k} = \mu + \alpha + \epsilon_{10k},$$

for the spleens of old animals,

$$X_{01k} = \mu - \alpha + \epsilon_{01k},$$

and for a mixture of p spleens from young animals and $(1 - p)$ spleens from old animals, where $0 \leq p \leq 1$,

$$X_{11k} = p(\mu + \alpha + (1 - p)(\mu - \alpha) - \gamma + \epsilon_{11k}$$
$$= \mu + (1 - 2p)(\alpha) - \gamma + \epsilon_{11k}$$

where the ϵ_{11k} are independent values.

Makinodan knew in advance of his experiment that the main effect α was nonzero. He also knew the distributions of the errors ϵ_{11k} would be different for the different populations. We can assume only that these errors are independent of one another and that their medians are zero.

Makinodan wanted to test the hypothesis that the interaction term $\gamma > 0$ because there are immediate biological interpretations for the three alternatives: from $\gamma = 0$ one may infer independent action of the two cell populations; $\gamma < 0$ means excess lymphocytes in young populations; and $\gamma > 0$ suggests the presence of suppressor cells in the spleens of older animals.

But what statistic are we to use to do the test? One possibility is

$$S = |\overline{X}_{11.} - p\overline{X}_{01.} - (1 - p)\overline{X}_{10.}|.$$

If the design were balanced or if we could be sure that the null effect $\mu = 0$, this is the statistic we would use. But the design is not balanced, with the result that the main effects (with which we are not interested) are confounded with the interaction (with which we are).[6]

Fortunately, another resampling method, the bootstrap, can provide a solution. Draw an observation at random and with replacement from the set $\{x_{10k}\}$; label it $x*_{10}$. Similarly, draw the bootstrap observations $x*_{01}$ and $x*_{11}$ from the sets $\{x_{01k}\}$ and $\{x_{11k}\}$. Let

$$\gamma* = px*_{01} + (1 - p)x*_{10} - x*_{11} \tag{8.8}$$

Repeat this resampling procedure several hundred times, obtaining a bootstrap estimate $\gamma*$ of the interaction each time you resample. Use the resultant set of bootstrap estimates $\{\gamma*\}$ to obtain a confidence interval for γ. If 0 belongs to

---

Mean DPFC response.

Effect of pooled old BC3FL spleen cells on the anti-SRBC response of indicator-pooled BC3FL spleen cells (data extracted from Makinodan et al. [1976]). Bootstrap analysis:

Young Cells	Old Cells	1/2 + 1/2
5,640	1,150	7,100
5,120	2,520	11,020
5,780	900	13,065
4,430	50	
7,230		

Bootstrap sample 1: $5,640 + 900 - 11,020 - 4,480$
Bootstrap sample 2: $5,780 + 1,150 - 11,020 - 4,090$
Bootstrap sample 3: $7,230 + 1,150 - 7,100 + 1,280$

. . . .
. . . .

Bootstrap sample 100: $5,780 + 2,520 - 7,100 + 1,200$

---

[6]The standard parametric (ANOVA) approach won't work in this example either.

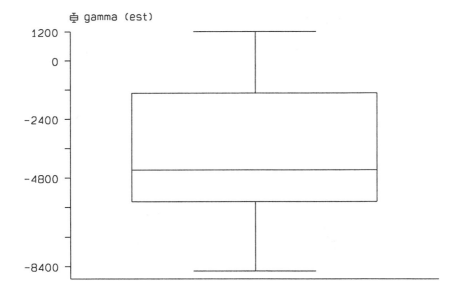

FIGURE 8.2. Box and whiskers plot of the bootstrap distribution of $\gamma*$.

this confidence interval, accept the hypothesis of additivity; otherwise reject. Our results are depicted in Figure 8.2.

The sample size in this experiment is small, and if we were to rely on this sample alone we could not rule out the possibility that $\gamma \geq 0$. Indeed, the 90% confidence interval stretches from $-7, 815$ to $+1, 060$. But Makinodan et al. [1976] conducted many replications of this experiment for varying values of p, with comparable results; they could feel confident in concluding $\gamma < 0$, showing that young spleens have an excess of lymphocytes.

8.7. Summary

In this chapter, you learned the principles of experimental design: to block or measure all factors under your control and to randomize with regard to factors that are not.

You learned to analyze balanced k-way designs for main effects, and you balanced two-way designs for both main effects and interactions. You learned to use the Latin square to reduce sample size and to use bootstrap methods when designs are not balanced.

8.8.   To Learn More

For more on the principles of experimental design, see Fisher [1935], Kempthorne [1955], Wilk and Kempthorne [1956, 1957], Scheffe [1959], Maxwell and Cole [1991]. Further sample-size guidelines are provided in Shuster [1993].

Permutation test have been applied to a wide variety of experimental designs including analysis of variance [Kempthorne 1955, 1966, 1969, 1977; Jin 1984; Soms 1985; Diggle, Lange, and Benes 1991; Ryan, Tracey, and Rounds 1996], clinical trials [Lachin 1988a,b], covariance [Oja 1987] crossovers [Shen and Quade 1986], factorial [Loughin and Noble 1997], growth curves [Foutz, Jensen, and Anderson 1985; Zerbe 1979a,b], matched pairs [Peritz 1985; Welch 1987, 1988; Rosenbaum 1988; Good 1991, Zumbo 1996], randomized blocks [Wilk 1955], restricted randomization [Smythe 1988], sequential clinical trials [Wei 1988; Wei, Smythe and Smith 1986; Lefebvre 1982]. Bradbury [1987] compares parametric with randomization tests. Mapleson [1986] applied the bootstrap to the analysis of clinical trials.

8.9.   Exercises

1. Design an experiment.

   a. List all the factors that might influence the outcome of your experiment.
   b. Write a model in terms of these factors.
   c. Which factors are under your control?
   d. Which of these factors will you use to restrict the scope of the experiment?
   e. Which of these factors will you use to block?
   f. Which of the remaining factors will you neglect initially, that is, lump into your error term?
   g. How will you deal with each of the remaining covariates?
   h. How many subjects/items will you observe in each subcategory?
   i. Write out two of the possible assignments of subjects to treatment.
   j. How many possible assignments are there in all?

2. A known standard was sent to six laboratories for testing.

   a. Are the results comparable among laboratories?

	Laboratory					
	A	B	C	D	E	F
1/1	221.1	208.8	211.1	208.3	221.1	224.2
1/2	224.2	206.9	198.4	214.1	208.8	206.9
1/3	217.8	205.9	213.0	209.1	211.1	198.4

   b. The standard was submitted to the same laboratories the following month. Are the results comparable from month to month?

	Laboratory					
	A	B	C	D	E	F
2/1	208.8	211.4	208.9	207.7	208.3	214.1
2/2	212.6	205.8	206.0	216.2	208.8	212.6
2/3	213.3	202.5	209.8	203.7	211.4	205.8

3. Potted tomato plants in groups of six were maintained in one of 3 levels of light and two levels of water. What effects, if any, did the different levels have on yield?

Light	Water	Yield	Light	Water	Yield
1	1	12	2	2	19
1	1	8	2	2	16
1	1	8	2	2	17
1	2	13	3	1	17
1	2	15	3	1	24
1	2	16	3	1	22
2	2	16	3	2	24
2	2	12	3	2	26
2	2	13	3	2	28

4. **a.** Are the two methods of randomization described in Section 8.1.3. equivalent?
   b. Suppose you had a six-sided die and three coins. How would you assign plots to one of four treatments? Three rows and two columns? Eight treatments?
5. Aflatoxin is a common and undesirable contaminant of peanut butter. Are there significant differences among the following brands?

Snoopy	0.5	7.3	1.1	2.7	5.5	4.3
Quick	2.5	1.8	3.6	5.2	1.2	0.7
Mrs. Good's	3.3	1.5	0.4	4.8	2.2	1.0

The aflatoxin content is given in parts per billion (ppb).
6. Four containers, each with 10 oysters, were randomly assigned to 4 stations, each kept at a different temperature in the wastewater canal of a power plant. The containers were weighed before treatment and after one month in the water. Were there statistically significant differences among the stations?

Trt	Initial	Final	Trt	Initial	Final
1	27.2	32.6	3	28.9	33.8
1	31.0	35.6	3	23.1	29.2
1	32.0	35.6	3	24.4	27.6
1	27.8	30.8	3	25.0	30.8
2	29.5	33.7	4	29.3	34.8
2	27.8	31.3	4	30.2	36.5
2	26.3	30.4	4	25.5	30.8
2	27.0	31.0	4	22.7	25.9

7. You can increase the power of a statistical test in three ways: a) making additional observations, b) making more precise observations, c) adding covariates. Discuss this remark in the light of your own experimental efforts.

8. A pregnant animal was accidentally exposed to a high dose of radioactivity. Her seven surviving offspring (one died at birth) were examined for defects. Three tissue samples were taken from each animal and examined by a pathologist. What is the sample size?

9. Show that if T' is a monotonic increasing function of T, that is, T' always increases when T increases, then a test based on the permutation distribution of T' will accept or reject only if a permutation test based on T also accepts or rejects.

10. Should your tax money be used to fund public television? Recall from Exercise 4.11 that when a random sample of adults were asked for their responses on a 9-point scale (1 is very favorable and 9 is totally opposed) the results were as follows:

   3, 4, 6, 2, 1, 1, 5, 7, 4, 3, 8, 7, 6, 9, 5.

   Turns out the first ten of these responses came from the City, and the last five came from the Burbs. Are there significant differences between the two areas? Given that two-thirds of the population live in the City, provide a point estimate and confidence interval for the mean response of the entire population. [*Hint*: Modify the test statistic so as to weight each block proportionately.]

11. Your company is considering making a takeover bid for a marginally profitable line of retail stores. Some of the stores are real moneymakers; others could be a major drain on your own company's balance sheet. You still haven't made up your mind, when you get hold of the following figures based on a sample of 15 of the stores:

Store Size	Percent Profit
Small	7.0, 7.3, 6.5, 6.4, 7.5
Medium	7.3, 8.0, 8.5, 8.2, 6.8
Large	8.7, 8.1, 8.9, 9.2, 9.5

   Have these figures helped you make up your mind? Write up a brief report for your CEO summarizing your analysis.

12. Using the insecticide data collected by Plackett and Hewlett [1963] in Table 8.6, test for independence of action using only the zero and highest dose level of the first drug. Can you devise a single test that would use the data from all dose levels simultaneously?

13. Take a third look at the experimental approaches described in Exercises 3.12 and 3.13. What might be the possible drawbacks of each? How would you improve on the methodology?

14. Do the four hospitals considered in Exercise 1.16 have the same billing practices? (*Hint*: Block the data before analyzing.)

15. In Exercise 5.9, we considered the effect of noise on productivity. The following table includes additional information concerning years of experience. Presumably, more experienced workers should be less bothered by noise. But is this the case?

Noise (db)	Duration (min)	Experience (yrs)	Noise (db)	Duration (min)	Experience (yrs)
0	11.5	4	40	16.0	8
0	9.5	7	60	20.0	5
20	12.5	3	60	19.5	7
20	13.0	2	80	28.5	8
40	15.0	11	80	27.5	10

16. In how many different ways can we assign 9 subjects to 3 treatments, given equal numbers in each sample? What if we began with 6 men and 3 women and wanted to block our experiment?

17. Unequal sample sizes? Take a second look at the data of Section 8.6.2. We seem to have three different samples with three different samples sizes, each drawn from a continuous domain of possible values. Or do we? From the viewpoint of the bootstrap, each sample represents our best guestimate of the composition of the larger population from which it was drawn. Each hypothetical population appears to consist of only a finite number of distinct values, a different number for each of the different populations. Discuss this seeming paradox.

CHAPTER 9

Multiple Variables and Multiple Hypotheses

The value of an analysis based on simultaneous observations of several variables such as height, weight, blood pressure, and cholesterol level, for example, is that it can be used to detect subtle changes that might not be detectable, except with very large, prohibitively expensive samples, were we to consider only one variable at a time.

Any of the resampling procedures can be applied in a multivariate setting provided we can find a single-valued test statistic that can stand in place of the multivalued vector of observations.

9.1. Hotelling's T^2

One example of such a statistic is Hotelling's T^2, a straightforward generalization of Student's t (discussed in Section 4.3.2) to the multivariate case. Suppose we have made a series of exchangeable vector-valued observations on J variables. In the ith vector of observations $\vec{X}_i = \{X_i^1, X_i^2, \ldots X_i^J\}$, X_i^1 might denote the height of the ith subject, X_i^2 his weight, and so on. Let $\vec{X}$ denote the vector of mean values $\{X^1, X^2, \ldots X^J\}$ and V the $J \times J$ covariance matrix; whose ijth element

---

Increase the Sensitivity of Your Experiments

- Take more observations.
- Take more precise observations.
- Block your experiments.
- Observe multiple variables.

V_{ij} is the covariance of X^i and X^j. To test the hypothesis that the midvalue of $\vec{X}_i = \vec{\xi}$, use[1] Hotelling's T^2

$$(\vec{X}_. - \vec{\xi})V^{-1}(\vec{X}_. - \vec{\xi}).$$

(note: the ith component of the vector $\vec{\xi}$ represents the hypothesized midvalue of the ith variable).

Loosely speaking, this statistic weighs the contribution of individual variables and pairs of variables in inverse proportion to their covariances. This has the effect of rescaling the contribution of each variable so that its contribution.

For the purpose of resampling, each vector of observations on an individual subject is treated as a single indivisible entity. When we relabel, we relabel on a subject-by-subject basis so that all observations on a single subject receive the same new label. If the original vector of observations on subject i consists of j distinct observations on j different variables $(x_i^1, x_i^2 \ldots x_i^j)$ and we give this vector a new label $\pi(i)$, then the individual observations remain together as a unit, each with the same new label $(x_{\pi(i)}^1, x_{\pi(i)}^2 \ldots x_{\pi(i)}^j)$.

9.2.   Two-Sample Multivariate Comparison

The two-sample multivariate comparison is only slightly more complicated. Let $n_1, \vec{X}_1; n_2, \vec{X}_2$ denote the sample size and vector of mean values of the first and second samples respectively. Under the null hypothesis, we assume that the two sets of vector-valued observations come from the same distribution (that is, the labels 1 and 2 are exchangeable between the samples). Let V denote the pooled estimate of the common covariance matrix; that is V_{ij} is the covariance estimate of X^i and X^j based on the combined sample.

$$(N-2)V_{ij} = \sum_{m=1}^{2} \sum_{k=1}^{n_m} (X_{mk}^i - X_{m.}^i)(X_{mk}^j - X_{m.}^j)$$

To test the hypothesis that the mid values of the two distributions are the same, we could use the statistic

$$T^2 = (\vec{X}_1 - \vec{X}_2)V^{-1}(\vec{X}_1 - \vec{X}_2)^T$$

but then we would be forced to recompute the covariance matrix V and its inverse V^{-1} for each new rearrangement. To reduce the number of computations, Wald and Wolfowitz [1943] suggest a slightly different statistic T' that is a monotonic[2]

---

[1] A definition of V^{-1} may be found in texts on matrix or vector algebra, but why reinvent the wheel? Routines for manipulating and in this case, inverting, matricies are readily available in Gauss, IMSL, SAS, S, and a number of C-libraries.

[2] A monotonic increasing function increases whenever its argument increases. An example of a monotonic increasing function is $f(x) = 32 + .92x$, another is $f(x) = \exp[x]$. A monotonic decreasing function decreases whenever its argument increases. If $x > 0$, then

---

Calculating the Wald-Wolfowitz variant of Hotelling's-T^2 Blood Chemistry Data from Warner et al. (X:1990)

ID	BC	Albumin	Uric Acid	Mean	Albumin	Uric Acid
2381	N	43	54	N	41.25	47.25
1610	N	41	33	Y	37.0	52.75
1149	N	39	50	Y–N	−4.25	7.50
2271	N	42	48			
1946	Y	35	72	C		
1797	Y	38	30		8.982	−21.071
575	Y	40	46		−21.071	197.571
39	Y	35	63			
				C^{-1}		
					.1487	.01594
					.01594	.006796

Hotelling's T^2
$$= (-4.257.50)C^{-1}(-4.257.50)^T$$
$$= 2.092$$

---

function of T.

$$\text{Let } U_j = \frac{1}{N} \sum_{i=1}^{2} \sum_{k=1}^{n_i} X_{ikj}$$

$$c_{ij} = \sum_{m=1}^{2} \sum_{k=1}^{n_i} (X_{mki} - U_i)(X_{mkj} - U_j).$$

Let C be the matrix whose components are the c_{ij}. Then

$$T'^2 = (\vec{X}_{1.} - \vec{X}_{2.})C^{-1}(\vec{X}_{1.} - \vec{X}_{2.})^T.$$

As with all permutation tests we proceed in three steps:

1. We compute the test statistic for the original observations.
2. We compute the test statistic for all relabelings.
3. We determine the percentage of relabelings that lead to values of the test statistic that are as or more extreme than the original value.

Hotelling's T^2 is the appropriate statistic to use if you suspect the data has a distribution close to that of the multivariate normal. Under the assumption of multivariate normality, the power of the permutation version of Hotelling's T^2 converges with increasing sample size to the power of the most powerful parametric test that is invariant under transformations of scale. As is the case with its univariate counterpart, Hotelling's T^2 is most powerful among unbiased tests for testing a

---

x^2 is a monotonic increasing function of x but x^2 is not a monotonic function for all x as it decreases for $x < 0$ until it reaches a minimum at 0 and then increases for $x > 0$.

general hypothesis $F = G$ against a specific normal alternative [Runger and Eaton, 1992].

While the stated significance level of the parametric version of Hotelling's T^2 can not be relied on for small samples if the data is not normally distributed [Davis, 1982], as always, the corresponding permutation test yields an exact significance level, provided the errors are exchangeable from sample to sample.

9.3.   The Generalized Quadratic Form

9.3.1.   Mantel's U

Mantel's U, $\Sigma\Sigma a_{ij}b_{ij}$ [Mantel 1967], is perhaps the most widely used of all multivariate statistics because of its broad range of applications.[3] In Mantel's original formulation, a_{ij} is a measure of the temporal distance between items i and j, while b_{ij} is a measure of the spatial distance. As an example, suppose t_i is the day on which the ith individual in a study came down with cholera and (x_i, y_i) denotes her position in space. For all i, j set $a_{ij} = 1/(t_i - t_j)$ and $b_{ij} = 1/\sqrt{(x_i - x_j)^2 + (y_i - y_j)^2}$.

A large value for U would support the view that cholera spreads by contagion from one household to the next. How large is large? As always, we compare the value of U for the original data with the values obtained when we fix the i's but permute the j's as in $U' = \Sigma\Sigma a_{ij}b_{i\pi[j]}$.

9.3.2.   An Example in Epidemiology

An ongoing fear of many parents is that something in their environment—asbestos or radon in the walls of their house, or toxic chemicals in their air and groundwater—will affect their offspring. Table 8.1 is extracted from data collected by Siemiatycki and McDonald [1972] on congenital neural-tube defects. Eyeballing the gradient along the diagonal of this table one might infer that births of ancephalic infants occur in clusters. We could test this hypothesis statistically

TABLE 9.1. Incidents of Pairs of Ancephalic Infants by Distance and Months Apart.

km Apart	< 1	< 2	< 4
< 1	39	101	235
< 5	53	156	364
< 25	211	652	1516

---

[3]By appropriately restricting the values of a_{ij} and b_{ij}, the definition of Mantel's U can be seen to include several of the standard measures of correlation, including those usually attributed to Pearson, Pitman, Kendall, and Spearman [Hubert, 1985].

using the methods of Chapter 7 for ordered categories, but a better approach, because the exact time and location of each event is known, is to use Mantel's U. The question arises as to which measures of distance and time we should use. Mantel [1967] reports striking differences between one analysis of epidemiologic data in which the coefficients are proportional to the differences in position and a second approach (which he recommends) to the same data in which the coefficients are proportional to the reciprocals of these differences.

Using Mantel's approach, a pair of infants born 5 kilometers and 3 months apart contribute $\frac{1}{3} * \frac{1}{5} = \frac{1}{15}$ to the statistic. Summing up the contributions from all pairs and repeating the summing process for a series of random rearrangements, Siemiatycki and McDonald conclude that the clustering of ancephalic infants is not statistically significant.

9.3.3.   Further Generalization

Mantel's U is quite general in its application. The coefficents need not correspond to space and time. In a completely disparate application in sociology [Hubert and Schultz, 1976], observers studied k distinct variables in each of a large number of subjects. Their object was to test a specific sociological model for the relationships among the variables. The $\{a_{ij}\}$ in Mantel's U were elements of the $k \times k$ sample-correlation matrix while the $\{b_{ij}\}$ were elements of an idealized or theoretical correlation matrix derived from the model. A large value of U supported the model; a small value would have ruled against it.

9.3.4.   The MRPP Statistic

The MRPP, or multiresponse permutation procedure [Mielke, 1979] has been applied to applications as diverse as the weather and the spatial distribution of archaeological artifacts. The MRPP uses the permutation-distribution of between-object distances to determine whether a classification structure has a nonrandom distribution in space or time.

An example of the application of the MRPP arises in the assignment of antiquities (artifacts) to specific classes based on their spatial locations in an archaeological dig. Presumably, the kitchen tools of primitive man—woks and Cuisinarts—should be found together, just as a future archaeologist can expect to find TV, VCR, and stereo side by side in a neolithic living room.

Following Berry et al. [1980, 1983], let $\Omega = \{\omega_1, \cdots, \omega_n\}$ designate a collection of N artifacts within a site; let $X_{i1} \cdots X_{ir}$ denote the r coordinates for the site space for the ith artifact; let $S_1, \cdots S_{g+1}$ represent an exhaustive partitioning of the N artifacts into $g+1$ disjoint classes, (the $g+1$st being reserved for not-yet-classified items); and let n_j be the number of artifacts in the jth class.

Define the distance between the ith and jth artifacts as

$$\delta_{ij} = 1/\sqrt{\sum_{k=1}^{r}(X_{ik} - X_{jk})^2}$$

Define the average between-artifact distance for all artifacts within the ith class,

$$\zeta_i = \frac{2}{n_i(n_i - 1)} \sum_{i<j} \delta_{ij}\phi_i[\omega_i]\phi_i[\omega_j]$$

where $\phi_i[\omega]$ is an indicator function that is 1 if $\omega \epsilon S_i$ and 0 otherwise.

The test statistic is the weighted within-class average of these distances,

$$\Delta = \frac{\sum_{i=1}^{g} n_i \zeta_i}{\sum_{i=1}^{g} n_i}$$

The permutation distribution associated with Δ is taken over all allocations of the N artifacts to the $g + 1$ classes with the same numbers of artifacts $\{n_1 \cdots n_{g+1}\}$ assigned to each class.

Sound complicated? Perhaps. But the complications are the result of an archeologist thinking in quantitative terms about a problem in archeology. The statistics are as straightforward as when we first discussed them in Chapter 3. Choose a test statistic, compute the statistic for the original observations, and determine the permutation distribution of the test statistic you've chosen.

9.4. Multiple Hypotheses

One of the difficulties with clinical trials and other large-scale studies is that frequently so many variables are under investigation that one or more of them is practically guaranteed to be significant by chance alone. If we perform 20 tests at the 5%, or 1/20, level, we expect at least one significant result on the average. If the variables are related (and in most large-scale medical and sociological studies the variables have complex interdependencies), the number of falsely significant results could be many times greater.

A resampling procedure outlined by Troendle [1995] allows us to work around the dependencies. Suppose we have measured k variables on each subject and are now confronted with k test statistics $s_1, s_2, \ldots s_k$. To make these statistics comparable, we need to standardize them and render them dimensionless, dividing each by its respective L_1- or L_2-norm. For example, if one variable, measured in centimeters, takes values like 144, 150, 156, and the other, measured in meters, takes values like 1.44, 1.50, and 1.56, we might set $t_1 = s_1/4$ and $t_2 = s_2/0.04$.

Next, we order the standardized statistics by magnitude so that $t_{(1)} \leq \cdots \leq t_{(k)}$. Denote the corresponding hypotheses as $H_{(1)}, \ldots, H_{(k)}$. The probability that at least one of these statistics will be significant by chance alone at the α level is $1 - (1 - \alpha)^k \approx k\alpha$. But once we have rejected one hypothesis (assuming it was false), there will only be $k - 1$ true hypotheses to guard against rejecting.

Begin with $i = 1$ and follow these steps:

1. Repeatedly resample the data (with or without replacement), estimating the cutoff value $\phi(\alpha, k - i + 1)$ such that $\alpha = \Pr\{T(k - i + 1) \le \phi(\alpha, k - i + 1)\}$ where $T(k - i + 1)$ is the largest of the $k - i + 1$ test statistics $t_{(1)} \cdots t_{(k-i+1)}$ for a given resample.
2. If $t_{(k-i+1)} \le \phi(\alpha, k - i + 1)$, then accept all the remaining hypotheses $H_{(1)} \ldots H_{(k-i+1)}$ and stop. Otherwise, reject $H_{(k-i+1)}$, increment i, and return to Step 1.

9.5.   Summary

In this chapter, you learned the essentials of multivariate analysis for one- and two-sample comparisons. You learned how to detect clustering in time and space and to validate clustering models. You used the generalized quadratic form in its several guises including Mantel's U and Mielke's multiresponse permutation procedure to work through applications in epidemiology and archeology. And you learned how to combine the results of multiple simultaneous analyses.

9.6.   To Learn More

Blair et al. [1994] and van-Putten [1987] review alternatives to Hotelling's T^2. Boyett and Shuster [1977] consider its medical applications. Extensions to other experimental designs are studied by Barton and David [1961].

Mantel's U has been rediscovered frequently, often without proper attribution (see Whaley, 1983). Empirical power comparisons between MRPP rank tests and with other rank tests are made by Tracy and Tajuddin [1985] and Tracy and Khan [1990].

The generalized quadratic form has seen widespread application in anthropology [Williams-Blangero 1989], archaeology [Klauber 1971, 75], ecology [Bryant, 1977; Douglas and Endler, 1982; Highton, 1977; Levin 1977; Mueller and Altenberg 1985; Royaltey, Astrachen and Sokal, 1975; Ryman et al., 1980; Syrjala 1996], earth science [Mieleke 1991], education [Schultz and Hubert, 1976], epidemiology [Alderson and Nayak, 1971; Fraumeni and Li, 1969; Glass and Mantel, 1969; Glass et al., 1971; Klauber and Mustacchi 1970; Kryscio et al., 1973; Mantel and Bailar, 1970; Merrington and Spicer, 1969; Siemiatycki and McDonald, 1972; Smith and Pike, 1976], geography [Cliff and Ord, 1971, 73, 81; Hubert, Golledge and Costanzo, 1981; Hubert et al. 1984], management science [Graves and Whinston, 1970], ornithology [Cade and Hoffman, 1993], paleontology [Marcus, 1969], psychology [Hubert and Schultz 1976], sociology [Hubert and Baker, 1977, 78], and systematics [Dietz, 1983; Gabriel and Sokal, 1969; Selander and Kaufman, 1975; Sokal, 1979]. Siemiatycki [1978] considered various refinements.

Blair, Troendle, and Beck [1996], Troendle [1995] and Westfall and Young [1989] expand on the use of permutation methods to analyze multiple hypotheses; also, see the earlier work of Shuster and Boyett [1979], Ingenbleek [1981] and Petrondas and Gabriel [1983]. Simultaneous comparsions in contingency tables are studied by Passing [1984]. For additional insight into the multivariate approach, including a discussion of resampling methods for analyzing repeated measures see Good [1994, Chapter 5] and Zerbe and Murphy [1986].

9.7. Exercises

1. You are studying a new tranquilizer that you hope will minimize the effects of stress. The peak effects of stress manifest themselves between 5 and 10 minutes after the stressful incident, depending on the individual. To be on the safe side, you've made observations at both the 5- and 10-minute marks.

Subject	Prestress	5-minute	10-minute	Treatment
A	9.3	11.7	10.5	Brand A
B	8.4	10.0	10.5	Brand A
C	7.8	10.4	9.0	Brand A
D	7.5	9.2	9.0	New drug
E	8.9	9.5	10.2	New drug
F	8.3	9.5	9.5	New drug

How would you correct for the prestress readings? Is this a univariate or a multivariate problem? List possible univariate and multivariate test statistics. Perform the permutation tests and compare the results.

2. Show that Pitman's correlation is a special case of Mantel's U.

CHAPTER 10

Classification and Discrimination

10.0. Introduction

Delivery of my third daughter Diana was prolonged, occasioned not by reluctance on my daughter's part, but by the need to assemble the delivery table. When we met up with the by-now-familiar obstetrician in the new wing of Bronson Methodist Hospital, he was staring at an assembly manual with a puzzled frown. "Not quite English," he said.

The parts of the table were stacked against the wall. While I unpacked and assembled them, Dr. Gershon read aloud from the manual. My wife, now my ex-wife, leaned against the door irritably and muttered something about "whenever you Bozos are ready."

We were ready, finally; my wife lay back on the table, and Diana emerged almost immediately. While the doctor fussed with his patient, I got to hold my daughter and record the first of her Apgar scores. The Apgar scoring system, used to assess whether a newborn is at risk, was detailed on a wall-chart, the delivery room's only decoration. Diana breathed normally, but her color wasn't quite pink enough. As I recall, I gave her an 8, which increased to a 10 when she was assessed a second time, 3 minutes into her life. This was my first, but not my last, experience with classification.

Is the patient John R. at risk? Should Martha be given an auto loan? What factors in Martha's past should we measure and how should we combine them? Is the image on the radar scope an enemy submarine or one of our own? Is this new compound likely to be a carcinogen? What and how many genera should new species be assigned to?

In this chapter, you'll resolve four separate problems related to classification and discrimination:

- Categorization—determining the number and extent of categories to which items and individuals should be assigned.
- Discrimination—deriving and applying a decision rule for the assignment of items and individuals to existing categories.
- Validation—using a newly gathered set of observations to validate an existing model.
- Cross-validation—resampling from the same data set used to derive a discrimination rule to validate the resulting model.

In addition to the by-now-familiar methods of bootstrap and permutation, you'll apply density estimation, nearest-neighbor, neural networks, and CART to voice-recognition, triage, credit-risk determination, and the assignment of plants and animals to species and genre.

10.1. Classification

The problem of classification is actually threefold:

1. Determine the number of subpopulations or categories into which a population should be divided.
2. Develop an optimal system for assigning patterns, items, and individuals to categories.
3. Validate this system.

The first of these is the most difficult and is still only partially resolved. We'll consider many of the issues and some partial solutions in the next few sections. Our focus is on what techniques to use and when to use them; because of the mathematical complexity of most, we leave the computational details to various commercially available computer programs.

Classification

Determine the number of categories:
     Bump-hunting 10.2.1.
     Block clustering 10.3.

Develop an assignment scheme:
     CART 10.4.
     Other methods 10.5.

Validate or cross-validate your results:
     10.6, 10.7.

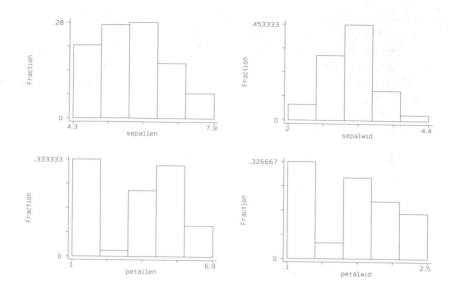

FIGURE 10.1. Sepal and Petal Measurements of 150 Iris Plants.

10.1.1.   A Bit of Detection

Let's begin with a simple example. The sepal length, sepal width, petal length, and petal width of 150 iris plants were recorded by Fisher [1936]. Figure 10.1 summarizes the findings in the form of histograms for each of the four variables.

Our objective, like Fisher's, is twofold: (1) to subdivide the sample of iris plants into k distinct species; and (2) to develop a methodology for classifying iris into species in the future.

Our first clues to the number of subpopulations or categories, as well as to the general shape of the underlying frequency distribution, come from consideration of the histogram (see Section 1.2.4).[1] A glance at two of the histograms in Figure 10.1 suggests the presence of at least two species, but because of the overlap of the various subpopulations it is difficult to be sure. *Density estimation* provides a technique by which we can make more precise subdivisions.

10.2.   Density Estimation

The histograms depicted in Figure 10.1 are defined over $m = 5$ non-overlapping intervals or bins, each of fixed length h, so our estimate of the underlying probability

---

[1] The histograms we constructed in Exercise 1.16 were used for just that purpose.

Which Kernel?

A variety of different kernels[2] have been proposed by various authors including:

$$K_G(z) = \frac{1}{\sqrt{2\pi}} e^{-z^2/2}$$

$$K_B(z) = \begin{cases} \frac{15}{16}(1 - z^2)^2 & \text{if } |z| < 1 \\ 0 & \text{otherwise} \end{cases}$$

$$K_C(z) = \begin{cases} 1 + \cos(2\pi z) & \text{if } |z| < 1/2 \\ 0 & \text{otherwise} \end{cases}$$

$$K_E(z) = \begin{cases} \frac{3}{4}(1 - \frac{1}{5}z^2)/\sqrt{5} & \text{if } |z| < \sqrt{5} \\ 0 & \text{otherwise} \end{cases}$$

$$K_P(z) = \begin{cases} \frac{4}{3} - 8z^2 + 8|z^3| & \text{if } |z| < 1/2 \\ \frac{8}{3}(1 - |z|)^3 & \text{otherwise} \end{cases}$$

$$K_T(z) = \begin{cases} \frac{15}{16}(1 - z^2)^2 & \text{if } |z| < 1 \\ 0 & \text{otherwise} \end{cases}$$

The choice of kernel K is not as important as the choice of the bandwidth or interval h.

---

density f may be written in the form of a step function:

$$f(x) = \frac{1}{hN} \sum_{i=1}^{5} N_i I[x_0 + (i - 1)h < x < x_0 + (i + 1)h]$$

where N_i denotes the number of observations in the ith interval, and the indicator function I is 1 if x lies in the indicated interval and 0 otherwise.

For estimating the density function, the histogram has several obvious limitations, including discontinuities at the boundaries of each cell and the fact that it is zero outside the minimum (x_0) and the maximum $(x_0 + mh)$ of the sample. We can smooth this function and eliminate the discontinuities by replacing the set of rectangles with a set of triangles as in the formula[3]

$$f(x) = \frac{1}{hN} \sum_{i=1}^{m} \frac{N_i(h/2 - |x - x_0 - ih + h/2|)}{h/2} I[x_0 + (i - 1)h < x < x_0 + ih]$$

But why stop with triangles? We could introduce a general smoothing kernel K, a function of $x - x_i$, so that

$$f(x) = \frac{1}{hN} \sum_{i=1}^{m} N_i K\left(\frac{x - x_i}{h}\right).$$

---

[3] $|x|$ denotes the absolute value of x. $|+3| = 3$ and $|-3| = 3$.

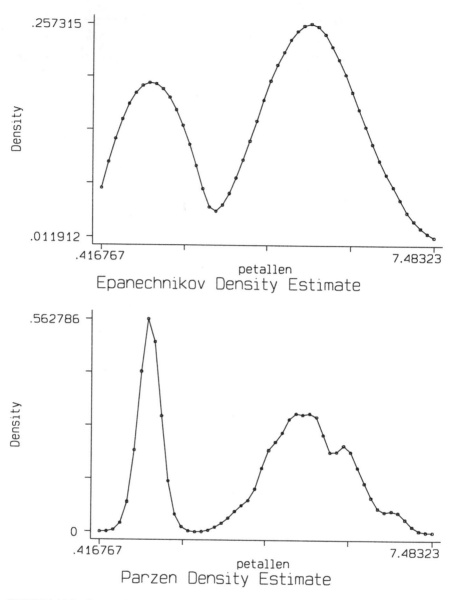

FIGURE 10.2. Comparison of Epanechnikov and Parzen Kernels Density Estimates of Distribution of Iris Petal Lengths.

The sidebar on the previous page lists some of the possible kernels that have been proposed over the years. As Figure 10.2 reveals, each type of kernel has its own distinct properties.

As can be seen in Figure 10.3, our choice of the bin width h is as critical as or more critical than our choice of kernel. Too few bins and we end up with

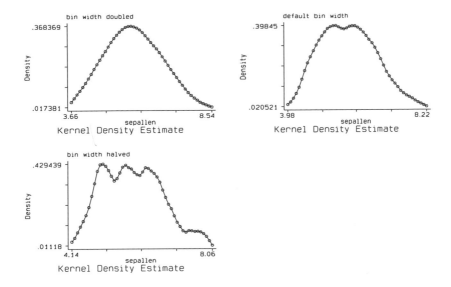

FIGURE 10.3. Effect of Bin Width on Epanechnikov Kernel Density Estimate of Distribution of Iris Petal Lengths.

a flat line; too many and our estimate has too many jumps. Or does it? If our purpose in observing petal length is to classify Iris into species, then the default bin width (upper right in Figure 10.3) suggests the presence of two distinct species; with bin width doubled (upper left) we conclude there is only one species, while with the bin width halved (lower left) it appears, there are as many as four.

The default value used by Stata has length $h = 0.9mn^{-1/5}$, where n is the sample size and m is the smaller of s, the standard deviation of the sample, and IQR, the interquartile range, the difference between the 75th and 25th percentiles, divided by 1.349. For a kernel density estimate of my class of 22 students, this default length is 14 cm.

Alternate approaches to bin-width determination include the leave-one-out (delete-1) and delete-50% methods of cross-validation, which we considered in Section 5.4, along with bootstrap selection; see Shao and Tu [1995] for details.

10.2.1.   More Detective Work

The density estimates suggest the presence of only two species, and the matrix of scatterplots in Figure 10.4 using both sepal and petal measurements would seem to confirm this number.

Figure 10.5 is a close-up of one of the scatterplots in Figure 10.4. The symbols 1, 2, and 3 represent the species of each plant—three species, not two. We completely failed to separate the two larger species using density estimation. Or did we? Take a

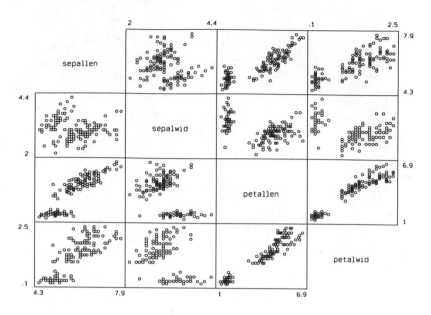

FIGURE 10.4. Sepal and Petal Measurements of 150 Iris Plants.

second look at Figure 10.2. The density estimate based on the Parzen kernel reveals three bumps. And that shoulder on the right end of the curve in the lower-left corner of Figure 10.3 corresponds to the third subpopulation.

10.2.2.   Bump Hunting

Three species of iris, not two! We should have seen it all along, and we would have if we'd just known where to look and which density estimate to rely on. But we didn't. Fortunately, there is a better way.

Recall again the series of kernel density estimators we constructed in Figure 10.3. Too few bins and the curve was unimodal; too many and distribution after distribution leaped out at us.

Four Steps to Classification

1. Use density estimation to fit a density curve to the data.
2. Determine the number of modes and thus, the number of subpopulations.
3. Fit a separate density curve to each of the subpopulations.
4. Use resampling to establish the optimal bin width.

Use these curves to classify any new observations.

Silverman [1981] capitalized on this observation, reasoning we could make any number of modes vanish if we just made the bin width h large enough. Suppose now we estimate the density using the formula

$$f[t, h, X] = \frac{1}{nh} \sum_{i=1}^{n} K\left(\frac{[t - X_i]}{h}\right)$$

and define h_k as the smallest value of h for which we still observe at most k modes.

Define $N(f)$ as the number of modes in the estimated distribution f. Then $N(f[., h_k, .]) = k$. To test the hypothesis H_k that the underlying density f has exactly k modes against the alternative K_{k+1} that it has more than k, we proceed as follows:

1. Draw n times without replacement from the original sample $\{X_i\}$ to get a bootstrap sample $\{X_i*\}$.
2. Smooth the bootstrap sample as in Section 4.7 by computing

$$Y_i* = c_k(X_i * + h_k Z_i) \text{ for } i = 1, 2, \ldots n$$

where the $\{Z_i\}$ are independent normal variables $N(0, 1)$ and the rescaling factor

$$c_k = (1 + \frac{h_k^2}{s^2})^{-1/2}$$

where s^2 is the sample variance of the original observations.
3. Form a density estimate $f(., h_k Y*)$ and count the number of modes.

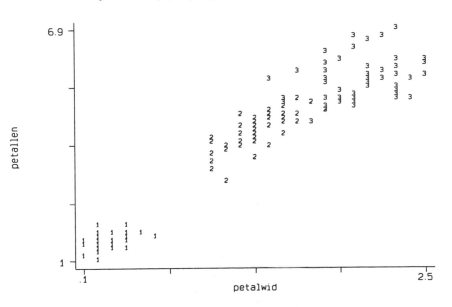

FIGURE 10.5. Petal Lengths and Widths of 150 Iris Plants

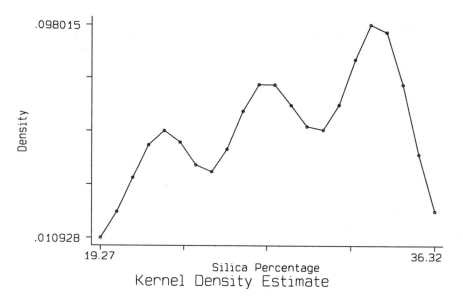

Kernel Density Estimate

FIGURE 10.6. Gaussian Kernel Density Estimate for Chondrite Data. $h = 3$.

TABLE 10.1. Percentages of Silica in 22 Chondrites; Ahrens [1965]

20.77,	22.56,	22.71,	22.99,	26.39,	27.08,	27.32,	27.33,	22.57,
27.81,	28.69,	29.36,	30.25,	31.69,	32.88,	33.23,	33.28,	33.40,
33.52,	33.83,	33.95,	34.82					

4. Repeat these initial steps 100 to 200 times and record the proportion of times P_k in which the number of modes exceeds k.

Silverman [1981] showed that if K is the Gaussian kernel, then $N(f_h) > k$ if and only if $h < h_k$. Applying these methods to the set of observations on chondrite meteors recorded in Table 10.1, Silverman [1981] obtained the results reproduced in Table 10.2, which suggest that the 22 chondrites came from three separate populations, a result illustrated in Figure 10.6.

TABLE 10.2. Critical Widths and Estimated Significance Levels for Chondrite Data.

Number of modes	Critical width	P_k
1	2.39	0.92
2	1.83	0.95
3	0.68	0.21
4	0.47	0.07

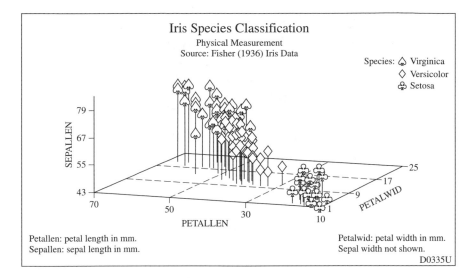

Iris Species Classification
Physical Measurement
Source: Fisher (1936) Iris Data

Species: ⚜ Virginica
◇ Versicolor
♣ Setosa

Petallen: petal length in mm.
Sepallen: sepal length in mm.

Petalwid: petal width in mm.
Sepal width not shown.

D0335U

FIGURE 10.7. Representing Three Variables in Two Dimensions; Iris Species Representation; Derived with the Help of SAS/GraphTM.

10.2.3. Detective Story: A 3-D Solution

We could apply Silverman's method to the Iris data (see Exercise 5), but another graphical method is faster. Taking advantage of SAS' 3-D graphing capabilities to plot sepal length and petal length and width simultaneously in Figure 10.7, the third species leaps out at us.

10.3. Block Clustering

Why didn't we use a three-dimensional graph in the first place? If we'd observed a dozen or more variables instead of merely two or three, would we have known which variables were relevant? And would we have known how to combine them?

A more formal and rewarding approach to classification is to set up a matrix whose rows represent the individuals (or objects) to be classified and whose columns stand for attributes of these same individuals. One such example, studied by Duffy, Fowlkes, and Kane [1987], was part of a strategic design effort for a telephone company. In their study, the data matrix consists of 99 rows corresponding to high-level functions such as "service-order formulation" and "vendor invoice processing" and 23 columns corresponding to objects like "customer," "circuit," and "service." The entries in their matrix are 1 or 0 according to whether the function associated with the row did or did not require information about the object associated with the column. An objective of their study is to identify and interpret clusters of functions with similar information needs, the "species" or building blocks of the telephone business.

TABLE 10.3. U.N. Votes by Country and Resolution

Country	K	K	C	P	SA	H	H	H	K	C
Albania	Y	Y	N	N	N	N	N	N	N	N
Brazil	Y	Y	N	Y	N	A	A	A	N	Y
Bulgaria	Y	N	Y	Y	A	Y	Y	Y	Y	N
France	Y	Y	Y	A	N	N	N	N	N	N
Kenya	Y	A	Y	Y	Y	N	N	N	A	N
Mexico	Y	Y	N	A	Y	A	A	A	N	Y
New Zealand	Y	Y	N	N	N	N	N	N	N	Y
Norway	Y	Y	Y	N	N	N	N	N	N	A
Senegal	Y	Y	A	Y	Y	N	N	N	A	A
Sweden	Y	Y	Y	N	N	N	N	N	N	N
Tanzania	Y	A	Y	Y	Y	A	A	A	A	N
USA	Y	Y	N	N	N	N	N	N	N	Y
USSR	Y	N	Y	Y	A	Y	Y	Y	Y	N
Venezuela	Y	Y	N	Y	Y	A	A	A	Y	N

Another example is depicted in Table 10.3a, whose rows correspond to countries and whose columns correspond to how these countries voted on various resolutions. The objective is to use the votes on U.N. resolutions to identify voting blocs within the international body. Preliminary examination of Table 10.3a reveals that the vote on the first Korean resolution (column 1), is completely uninformative and may be discarded from further consideration. While no other obvious distinctions suggest themselves, rearranging rows as in Table 10.3b reveals a definite separation. The three Hungarian resolutions, columns 7, 8, and 9 (actually three amendments to a resolution on South Africa proposed by Hungary), clearly divide the U.N. into three distinct voting blocs.

Is this division statistically significant? The answer, due to Duffy and Quinoz [1991], requires three steps. First, we need to establish a measure of heterogeneity, for example, the reduction in sums of squares about the mean attained by dividing the rows of the table into three blocks and computing three separate sums. Table 10.3b represents the maximum reduction possible for the data of Table 10.3a. Next, we need to rearrange the data among the columns and the rows (so that the numbers of *Nos*, *Yess* and *Abstains* remain constant) and again determine the maximum reduction possible with a division into three blocks. Proceeding in this fashion, we obtain a distribution of the possible values of the reduction statistic against which the reduction accomplished by our original partition may be compared.

10.4.   CART

As you've seen in the preceding sections, classification is a difficult and challenging problem. The task of *triage*, or *discrimination*, is a great deal simpler, *if* we have available to us a training set of already classified items.

TABLE 10.4. U.N. Votes by Country and Resolution Rows Rearranged

Country	K	K	C	P	SA	H	H	H	K	C
Albania	Y	Y	N	N	N	N	N	N	N	N
France	Y	Y	Y	A	N	N	N	N	N	N
Kenya	Y	A	Y	Y	Y	N	N	N	A	N
New Zealand	Y	Y	N	N	N	N	N	N	N	Y
Norway	Y	Y	Y	N	N	N	N	N	N	A
Senegal	Y	Y	A	Y	Y	N	N	N	A	A
Sweden	Y	Y	Y	N	N	N	N	N	N	N
USA	Y	Y	N	N	N	N	N	N	N	Y
Brazil	Y	Y	N	Y	N	A	A	A	N	Y
Mexico	Y	Y	N	A	Y	A	A	A	N	Y
Tanzania	Y	A	Y	Y	Y	A	A	A	A	N
Venezuela	Y	Y	N	Y	Y	A	A	A	Y	N
USSR	Y	N	Y	Y	A	Y	Y	Y	Y	N
Bulgaria	Y	N	Y	Y	A	Y	Y	Y	Y	N

CART^{TM} (Computer-Aided Binary Regression Tree), developed by Breiman et al. [1984] provides a systematic, more powerful way to explore and use our observations.

This method of *discrimination* subdivides a set of observations into k distinct predefined classes based on the values of M categorical or ordinal variables. The idea is to pose a series of questions to a *training set* whose observations are already categorized. Each question subdivides the *training set* (or a subset thereof) into two groups of more homogeneous composition. These questions may be either simple:

Is the kth variable large?

or complex:

Is the sum of the jth and kth variables large while the mth variable is small?

Applying this same approach to the Iris data we considered previously and proceeding one variable at a time, CART^{TM} constructs a tree consisting of two nodes. First, CART establishes that petal length is the most informative of the four variables and the tree is split on petal length less than or equal to 2.450. Node 2 is split on petal width less than or equal to 1.750. CART concludes that the remaining variables do not contribute significant additional information.

We have 96% correctly classified! (See Table 10.4.) A moment's thought dulls the excitement; this is the same set of observations we used to develop our selection criteria. Of course, we were successful. It's our success with new plants whose species we don't know in advance that is the true test.

Not having a set of yet-to-be-classified plants on hand, we make use of the existing data as follows: Divide the original group of 150 plants into 10 subgroups, each consisting of 15 plants. Apply the binary-tree-fitting procedure 10 times

TABLE 10.5. Learning Sample Classification Table.

Actual Class	Predicted Class 1	2	3	Actual Total
1	50.000	0.000	0.000	50.000
2	0.000	49.000	1.000	50.000
3	0.000	5.000	45.000	50.000
Pred. Tot.	50.000	54.000	46.000	150.000
Correct	1.000	0.980	0.900	
Success Ind.	0.667	0.647	0.567	
Tot. Correct	0.960			

in succession, each time omitting one of the 10 groups, each time applying the tree it developed to the omitted group. Only 10 of the 150 plants, or 7.5%, were missclassified (Table 10.5), a 92.5% success rate, suggesting that in future samples consisting of Iris of unknown species, the system of classification developed here would be successful 92.5% of the time.

10.4.1. Refining the Discrimination Scheme

CART has three other features that command our attention. First, it is parsimonious, that is, all other factors being equal, it is programmed to produce a tree with the fewest number of branches. Only two of the four variables were used in the current example.

Second, CART may be programmed to take account of the expected composition of the target population. If we know we'll be searching in an area where species 3 comprises almost 50% of the population (in contrast to the training set in which species 3 is relatively rare), we can specify a high a priori probability for species 3 when setting up the tree-building procedure.

Suppose we are using CART to help us appraise a set of credit applications. We might have in mind three categories: 1) grant credit, 2) request additional information, and 3) reject application. Quite different costs are associated with each form of misclassification. We can specify these costs when setting up the tree-building procedure, ensuring that their relative values will be accounted for in the discrimination rule.

TABLE 10.6. Cross-Validation Classification Table.

Class	Prior Prob.	├——Cross Validation——┤ N	N Mis-Classified	Cost	├——Learning Sample——┤ N Mis-N	Classified	Cost
1	0.333	50	0	0.000	50	0	0.000
2	0.333	50	5	0.100	50	1	0.020
3	0.333	50	5	0.100	50	5	0.100
Tot	1.000	150	10		150	6	

10.5. Other Discrimination Methods

Two other promising methods of classification deserve mention: *k-point nearest neighbor* classification, proposed by Fix and Hodges [1951], yields a set of density estimates. A sphere is chosen around each point x that is big enough to contain k points in the training set. Then the estimate of the density of the ith class at x is

$$\hat{p}_i(x) = \frac{\{\text{number of points belonging to the } i\text{th class}\}}{\{\text{volume of the sphere}\}}$$

Neural nets and their relation to other methods of classification are surveyed in Ripley [1993]. Roeder [1990] proposed a set of density estimators of form

$$\hat{p}_i(x) = \Sigma w_{ij} f_j(x)$$

where the $\{f_j\}$ are a set of distributions of known form, say a set of normal distributions, and the weights $\{w_{ij}\}$ are determined by a neural network.

10.6. Validation

Validating an existing model, whether one of classification or regression, is the easiest of the problems we confront in this chapter as we can take advantage of techniques with which we are already familiar.

As a simple example, suppose the hypothesis to be tested is that certain events (births, deaths, accidents) occur randomly over a given time interval. If we divide this time interval into m equal parts and let p_k denote the probability of an event in the kth subinterval, the null hypothesis becomes $H : p_k = 1/m$ for $k = 1 \ldots m$. Suppose f_k is the number of critical events we observe in the ith subinterval. We express our results in the form of a $1 \times m$ contingency table:

Interval	1	2	3	$\cdots$	m
Observed	f_1	f_2	f_3	$\cdots$	f_m

Our test statistic is

$$\chi^2 = \sum_{k=1}^{m}(f_k - \frac{1}{m})^2.$$

To determine whether the observed value of this test statistic χ^2 is large, small, or merely average, we examine its distribution for all sets of frequencies $\{f*_k\}$ for which $\sum_{k=1}^{m} f*_k = \sum_{k=1}^{m} f_k$. We reject the hypothesis if only a small fraction of such sets have more extreme values of χ^2.

We can obtain a still more powerful test if we know more about the underlying model and can focus on a narrower class of alternatives.

Suppose we use the m categories to record the results of many repetitions of a series of $m - 1$ binomial trials, that is, we let the ith category correspond to the number of repetitions that result in exactly $i - 1$ successes. If our hypothesis is that the probability of success is .5 in each individual trial, then the expected number of repetitions resulting in exactly k successes is

$$\pi_k[.5] = n\binom{m}{k}(.5)^m.$$

If we proceed as we did in the preceding example, then our test statistic would be

$$S_1 = \sum_{k=1}^{m} \frac{(f_k - \pi_k[.5])^2}{\pi_k[.5]}.$$

Such a test provides us with protection against a wide variety of alternatives. But from the description of the problem we see that we can restrict ourselves to alternatives for which

$$\pi_k[p] = n\binom{m}{k}(p)^m(1-p)^{m-k}.$$

Fix, Hodges, and Lehmann [1959] show that a more powerful test statistic against such alternatives is $S = S_1 - S_2$ where

$$S_2 = \min_p \sum_{k=1}^{m} \frac{(f_k - \pi_k[p])^2}{\pi_k[p]}.$$

The permutation distribution of S is readily computed.

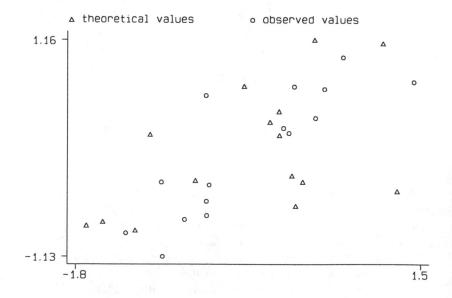

FIGURE 10.8. Comparison of Theoretical and Observed Values.

10.6.1.   A Multidimensional Model

With the availability of more powerful computers, scientists are addressing problems involving hundreds of variables, such as weather prediction and economic modeling. In many instances, the computer has taken the place of a laboratory. Where else could you emulate 1,000 years of stellar evolution or repeatedly study the effects of turbulence on an aircraft landing? Whether the data is real or simulated, we need to analyze our results. Figure 10.9 depicts the typical situation. The circles correspond to a set of points generated in a computer simulation, the xs to a set of observed values. Are the xs close enough to their predicted values that we may conclude our model is correct?

To distinguish between close and distant, we first need a *metric* and then a range of typical values for that metric. A metric m defined on two points x and y has the following properties:

$$m(x, y) \geq 0$$
$$m(x, x) = 0$$
$$m(x, z) \leq m(x, y) + m(y, z).$$

A good example is the standard Euclidian metric we use to measure the distance between two points x, y whose Cartesian coordinates are (x_1, x_2, x_3) and (y_1, y_2, y_3)

$$E(x, y) = \sqrt{(x_1 - y_1)^2 + (x_2 - y_2)^2 + (x_3 - y_3)^2}.$$

This metric can be applied even when x_i is not a coordinate in space but is the value of some variable such as blood pressure, laminar flow, or return on equity; it can easily be extended to any number of dimensions n as in

$$\sqrt{(x_1 - y_1)^2 + (x_2 - y_2)^2 + \cdots + (x_n - y_n)^2}.$$

The next step is to establish the range of possible values that such a metric might take by chance, using either the bootstrap or permutations of the combined set of theoretical and observed values. In Figure 10.9, for example, we might develop a metric as follows: For each point in the set of theoretical values, record the number of points among its three nearest neighbors (nearest in terms of the Euclidean metric), which are also in the set of theoretical values; sum the number of points so recorded. Call this sum S_0. S_0 equals 26 in the current case.

To determine whether S_0 is an extreme value, compute S for 50 to 100 random rearrangements of the combined group of theoretical and observed values. In my simulations, S ranged from 10 to 50 and I concluded that the theoretical model provided a reasonable explanation for the values actually observed.

Selecting a Metric: Structured Exploratory Data Analysis

Karlin and Williams [1984] use permutation methods in a structured exploratory data analysis (SEDA) of familial traits. A SEDA has four principal steps:

1. The data are examined for heterogeneity, discreteness, outliers, and so on, after which they may be adjusted for covariates (as in Chapter 5) and the appropriate transform applied (as in Chapter 3).
2. A collection of summary SEDA statistics are formed from ratios of functionals.
3. The SEDA statistics are computed for the original family trait values and for reconstructed family sets formed by permuting the trait values within or across families.
4. The values of the SEDA statistics for the original data are compared with the resulting permutation distributions.

As one example of a SEDA statistic, consider the OBP, the offspring-between-parent SEDA statistic:

$$\frac{\sum_i^N \sum_j^{J_i} |O_{ij} - (M_i + F_i)/2|}{\sum_i^N |F_i - M_i|}.$$

In family $i = 1, \ldots I$, F_i and M_i are trait values of the father and mother (cholesterol levels in their blood, for example), while O_{ij} is the corresponding trait value of the jth child of those same parents, $j = 1, \ldots, J_i$.

To evaluate the permutation distribution of the OBP, they consider all permutations in which the children are kept together in their respective family units, while we either (a) randomly assign to them a father and (separately) a mother, or (b) randomly assign to them an existing pair of spouses. The second of these methods preserves the spousal interaction. Which method one chooses will depend on the alternative(s) of interest.

To obtain the permutation distribution for the OBP statistic, substitute its formula in our sample programs in Appendices 1 and 2.

We see from this example that a validation metric specific to the problem can almost always be derived.

10.7.   Cross-Validation

The validation techniques outlined in the preceding section are appropriate when the model is already in existence, deduced from other data sets or sprung forth full-born, the product of a year spent lolling about in the public baths (Archimedes) or girlie shows (Feynman). But in many cases—the Iris and Chondrite data are examples—we are forced to make use of the same data to both develop the model and validate it.

Four techniques of cross-validation are in general use:

1. *K-fold*, the technique used by CART, in which we subdivide the data into K roughly equal-sized parts, then repeat the modeling process K times, leaving one section out each time for validation purposes,
2. *Leave-one-out*, an extreme example of K-fold, in which we subdivide into as many parts as there are observations. We leave one observation out of our clas-

sification procedure and use the remaining $n - 1$ observations as a training set. Repeating this procedure n times, omitting a different observation each time, we arrive at a figure for the number and percentage of observations classified correctly. A method that requires this much computation would have been unthinkable before the advent of inexpensive readily available high-speed computers. Today, at worst, we need step out for a cup of coffee while our desktop completes its efforts.

3. *Jackknife*, an obvious generalization of the leave-one-out approach, where the number left out can range from one observation to half the sample.
4. *Delete $- d$*, studied in Section 5.5, where we set aside a random percentage d of the observations for validation purposes, use the remaining $100 - d\%$ as a training set, then average over 100 to 200 such independent random samples. Shao and Tu [1995; p.306–313] show that the delete $- 50\%$ method is far superior to delete $- 1$.

The bootstrap (Chapter 5) is a fifth, recommended alternative [Efron 1983].

10.8.   Summary and Further Readings

In this chapter, we distinguished among classification, validation, and cross-validation and surveyed some of the many methods including density estimation, block clustering, CART, and restricted χ^2.

Rosenblatt [1956] and Parzan [1961] introduce density estimation, while the work of Silverman [1978] and Scott et al. [1978] are among its first practical applications. Applications are reviewed in Fryer [1976], Wertz [1978], Aitchison and Lauder [1985], Silverman [1986], Izenman and Sommer [1988], Fukunaga [1990] and Izenman [1991]. Multivariate applications are considered by Thompson

---

The Ad-Hoc, Post-Hoc Fallacy

Scientists inspired by a set of observations should and will formulate hypotheses. But it is a major error for those scientists to rely on that same set of data as "proof" of the hypotheses.

A frequent and flagrant example of the ad-hoc, post-hoc fallacy at work is the sociologist who, observing an unexpected difference between men and women (blacks and whites, Hispanics and Asians), then performs a one-tailed test on the data. Absent a preconceived hypothesis, the only appropriate test is two-tailed.

A scientist who gathers a large amount of information about a small group of subjects must anticipate that 5% (1 in 20) of the statistical tests at the 5% level will yield a statistically significant result as a result of a Type I error. One's first set of observations must be viewed as exploratory. Having found statistically significant results, the scientist should perform a second study with a new group of subjects, focusing on those variables found to be significant in the first study.

and Scott [1983], Thompson and Tapia [1990] and Scott [1991]. Ahmad and Kochar [1989] and Aly [1990] use density estimation for hypothesis testing, Izenman [1991] for comparing two populations.

For advice on bandwidth selection, see Abramson [1982a,b] and Park and Maron [1990]; for determining the optimal number of clusters, see Peck, Fisher, and Van Ness [1989]. Devroye and Gyorfi [1985] propose using the L_1 metric.

The chondrite data were also studied by Leonard [1978] and Good and Gaskins [1980]. This latter paper is accompanied by an extensive and critical discussion.

The resampling approach to validation has been used in DNA sequencing (Doolittle [1981] and Karlin et al. [1983] adopt permutation methods while Hasegawa, Krishino and Yano [1988] apply the bootstrap), ecology [Solow 1990], medicine [Arndt et al. 1996; Bullmore et al. 1996; Thompson et al. 1981; Titterington et al. 1981], nuclear power generation [Dubuisson and Lavison 1980], physics [Priesendorfer and Barnett 1983; DeJager, Swanepoel and Raubenheimer 1986], archeology [Berry et al 1980], business administration [Carter and Catlett 1987; Chhikara 1989; Thompson, Bridges and Ensor 1992] and chemistry [Penninckz et al. 1996].

The method of block classification was introduced by Hartigan [1972,1975]. The first use of resampling methods in classification was by Hubert and Levin [1976] and Hubert and Baker [1977].

CART methodology is due to Breiman et al. [1984]. The use of cross-validation is studied by Breiman [1992] and Shao and Tu [1995]. Mielke [1986] reviews alternate metrics.

Nearest-neighbor analysis is applied by Cover and Hart [1967] and Henze [1988]. Marron [1987], Hjorth [1994], Shao and Tu [1995; p306-313], and Stone [1974] provide a survey of parameter selection and cross-validation methods. The ad-hoc, post-hoc fallacy is further illustrated in the writings of Diaconis [1976].

10.9. Exercises

1. Using the Gaussian Kernel, adjust the bin width so the chondrite data displays 1, 2, 3, and 3+ modes.
2. Apply density estimation techniques to a data set your instructor provides or to one of the larger (100+ observations) data sets included with one of the commercially available statistics programs. Graph the data before you begin, then experiment with different kernels and different bandwidths. Take the square of each of the observations and repeat the process.
3. Homer and Lemeshow [1989] provide data on 189 births at a U.S. hospital. The last value in each row in the following table is the birthweight in grams. The data may be downloaded from **users.oco.net/drphilgood**.

   a. Use density estimation to determine whether this population should be subdivided.

b. How would you go about predicting a low birthweight (defined as less than 2.5 kilograms)? Reading across each row the variables are observation number, low birthweight (no 0, yes 1)?, age of mother in years, mother's weight in pounds at last menstrual period, race (white, black, other), smoked during pregnancy (no 0, yes 1)?, number of previous premature labors, history of hypertension (no 0, yes 1)?, has uterine irritability (no 0, yes 1)?, number of physician visits in the first trimester, and birthweight in grams.

4,1,28,120,3,1,1,0,1,0,709	10,1,29,130,1,0,0,0,1,2,1021
11,1,34,187,2,1,0,1,0,0,1135	13,1,25,105,3,0,1,1,0,0,1330
15,1,25,85,3,0,0,0,1,0,1474	16,1,27,150,3,0,0,0,0,0,1588
17,1,23,97,3,0,0,0,1,1,1588	18,1,24,128,2,0,1,0,0,1,1701
19,1,24,132,3,0,0,1,0,0,1729	20,1,21,165,1,1,0,1,0,1,1790
22,1,32,105,1,1,0,0,0,0,1818	23,1,19,91,1,1,2,0,1,0,1885
24,1,25,115,3,0,0,0,0,0,1893	25,1,16,130,3,0,0,0,0,1,1899
26,1,25,92,1,1,0,0,0,0,1928	27,1,20,150,1,1,0,0,0,2,1928
28,1,21,200,2,0,0,0,1,2,1928	29,1,24,155,1,1,1,0,0,0,1936
30,1,21,103,3,0,0,0,0,0,1970	31,1,20,125,3,0,0,0,1,0,2055
32,1,25,89,3,0,2,0,0,1,2055	33,1,19,102,1,0,0,0,0,2,2082
34,1,19,112,1,1,0,0,1,0,2084	35,1,26,117,1,1,1,0,0,0,2084
36,1,24,138,1,0,0,0,0,0,2100	37,1,17,130,3,1,1,0,1,0,2125
40,1,20,120,2,1,0,0,0,3,2126	42,1,22,130,1,1,1,0,1,1,2187
43,1,27,130,2,0,0,0,1,0,2187	44,1,20,80,3,1,0,0,1,0,2211
45,1,17,110,1,1,0,0,0,0,2225	46,1,25,105,3,0,1,0,0,1,2240
47,1,20,109,3,0,0,0,0,0,2240	49,1,18,148,3,0,0,0,0,0,2282
50,1,18,110,2,1,1,0,0,0,2296	51,1,20,121,1,1,1,0,1,0,2296
52,1,21,100,3,0,1,0,0,4,2301	54,1,26,96,3,0,0,0,0,0,2325
56,1,31,102,1,1,1,0,0,1,2353	57,1,15,110,1,0,0,0,0,0,2353
59,1,23,187,2,1,0,0,0,1,2367	60,1,20,122,2,1,0,0,0,0,2381
61,1,24,105,2,1,0,0,0,0,2381	62,1,15,115,3,0,0,0,1,0,2381
63,1,23,120,3,0,0,0,0,0,2410	65,1,30,142,1,1,1,0,0,0,2410
67,1,22,130,1,1,0,0,0,1,2410	68,1,17,120,1,1,0,0,0,3,2414
69,1,23,110,1,1,1,0,0,0,2424	71,1,17,120,2,0,0,0,0,2,2438
75,1,26,154,3,0,1,1,0,1,2442	76,1,20,105,3,0,0,0,0,3,2450
77,1,26,190,1,1,0,0,0,0,2466	78,1,14,101,3,1,1,0,0,0,2466
79,1,28,95,1,1,0,0,0,2,2466	81,1,14,100,3,0,0,0,0,2,2495
82,1,23,94,3,1,0,0,0,0,2495	83,1,17,142,2,0,0,1,0,0,2495
84,1,21,130,1,1,0,1,0,3,2495	85,0,19,182,2,0,0,0,1,0,2523
86,0,33,155,3,0,0,0,0,3,2551	87,0,20,105,1,1,0,0,0,1,2557
88,0,21,108,1,1,0,0,1,2,2594	89,0,18,107,1,1,0,0,1,0,2600
91,0,21,124,3,0,0,0,0,0,2622	92,0,22,118,1,0,0,0,0,1,2637
93,0,17,103,3,0,0,0,0,1,2637	94,0,29,123,1,1,0,0,0,1,2663
95,0,26,113,1,1,0,0,0,0,2665	96,0,19,95,3,0,0,0,0,0,2722
97,0,19,150,3,0,0,0,0,1,2733	98,0,22,95,3,0,0,1,0,0,2751
99,0,30,107,3,0,1,0,1,2,2750	100,0,18,100,1,1,0,0,0,0,2769
101,0,18,100,1,1,0,0,0,0,2769	102,0,15,98,2,0,0,0,0,0,2778
103,0,25,118,1,1,0,0,0,3,2782	104,0,20,120,3,0,0,0,1,0,2807

105,0,28,120,1,1,0,0,0,1,2821   106,0,32,121,3,0,0,0,0,2,2835
107,0,31,100,1,0,0,0,1,3,2835   108,0,36,202,1,0,0,0,0,1,2836
109,0,28,120,3,0,0,0,0,0,2863   111,0,25,120,3,0,0,0,1,2,2877
112,0,28,167,1,0,0,0,0,0,2877   113,0,17,122,1,1,0,0,0,0,2906
114,0,29,150,1,0,0,0,0,2,2920   115,0,26,168,2,1,0,0,0,0,2920
116,0,17,113,2,0,0,0,0,1,2920   117,0,17,113,2,0,0,0,0,1,2920
118,0,24,90,1,1,1,0,0,1,2948   119,0,35,121,2,1,1,0,0,1,2948
120,0,25,155,1,0,0,0,0,1,2977   121,0,25,125,2,0,0,0,0,0,2977
123,0,29,140,1,1,0,0,0,2,2977   124,0,19,138,1,1,0,0,0,2,2977
125,0,27,124,1,1,0,0,0,0,2922   126,0,31,215,1,1,0,0,0,2,3005
127,0,33,109,1,1,0,0,0,1,3033   128,0,21,185,2,1,0,0,0,2,3042
129,0,19,189,1,0,0,0,0,2,3062   130,0,23,130,2,0,0,0,0,1,3062
131,0,21,160,1,0,0,0,0,0,3062   132,0,18,90,1,1,0,0,1,0,3062
133,0,18,90,1,1,0,0,1,0,3062   34,0,32,132,1,0,0,0,0,4,3080
135,0,19,132,3,0,0,0,0,0,3090   136,0,24,115,1,0,0,0,0,2,3090
137,0,22,85,3,1,0,0,0,0,3090   138,0,22,120,1,0,0,1,0,1,3100
139,0,23,128,3,0,0,0,0,0,3104   140,0,22,130,1,1,0,0,0,0,3132
141,0,30,95,1,1,0,0,0,2,3147   42,0,19,115,3,0,0,0,0,0,3175
143,0,16,110,3,0,0,0,0,0,3175   144,0,21,110,3,1,0,0,1,0,3203
145,0,30,153,3,0,0,0,0,0,3203   146,0,20,103,3,0,0,0,0,0,3203
147,0,17,119,3,0,0,0,0,0,3225   148,0,17,119,3,0,0,0,0,0,3225
149,0,23,119,3,0,0,0,0,2,3232   150,0,24,110,3,0,0,0,0,0,3232
151,0,28,140,1,0,0,0,0,0,3234   154,0,26,133,3,1,2,0,0,0,3260
155,0,20,169,3,0,1,0,1,1,3274   56,0,24,115,3,0,0,0,0,2,3274
159,0,28,250,3,1,0,0,0,6,3303   160,0,20,141,1,0,2,0,1,1,3317
161,0,22,158,2,0,1,0,0,2,3317   162,0,22,112,1,1,2,0,0,0,3317
163,0,31,150,3,1,0,0,0,2,3321   164,0,23,115,3,1,0,0,0,1,3331
166,0,16,112,2,0,0,0,0,0,3374   167,0,16,135,1,1,0,0,0,0,3374
168,0,18,229,2,0,0,0,0,0,3402   169,0,25,140,1,0,0,0,0,1,3416
170,0,32,134,1,1,1,0,0,4,3430   172,0,20,121,2,1,0,0,0,0,3444
173,0,23,190,1,0,0,0,0,0,3459   174,0,22,131,1,0,0,0,0,1,3460
175,0,32,170,1,0,0,0,0,0,3473   176,0,30,110,3,0,0,0,0,0,3544
177,0,20,127,3,0,0,0,0,0,3487   179,0,23,123,3,0,0,0,0,0,3544
180,0,17,120,3,1,0,0,0,0,3572   181,0,19,105,3,0,0,0,0,0,3572
182,0,23,130,1,0,0,0,0,0,3586   183,0,36,175,1,0,0,0,0,0,3600
184,0,22,125,1,0,0,0,0,1,3614   185,0,24,133,1,0,0,0,0,0,3614
186,0,21,134,3,0,0,0,0,2,3629   187,0,19,235,1,1,0,1,0,0,3629
188,0,25,95,1,1,3,0,1,0,3637   189,0,16,135,1,1,0,0,0,0,3643
190,0,29,135,1,0,0,0,0,1,3651   191,0,29,154,1,0,0,0,0,1,3651
192,0,19,147,1,1,0,0,0,0,3651   193,0,19,147,1,1,0,0,0,0,3651
195,0,30,137,1,0,0,0,0,1,3699   196,0,24,110,1,0,0,0,0,1,3728
197,0,19,184,1,1,0,1,0,0,3756   199,0,24,110,3,0,1,0,0,0,3770
200,0,23,110,1,0,0,0,0,1,3770   201,0,20,120,3,0,0,0,0,0,3770
202,0,25,241,2,0,0,1,0,0,3790   203,0,30,112,1,0,0,0,0,1,3799
204,0,22,169,1,0,0,0,0,0,3827   205,0,18,120,1,1,0,0,0,2,3856
206,0,16,170,2,0,0,0,0,4,3860   207,0,32,186,1,0,0,0,0,2,3860
208,0,18,120,3,0,0,0,0,1,3884   209,0,29,130,1,1,0,0,0,2,3884
210,0,33,117,1,0,0,0,1,1,3912   211,0,20,170,1,1,0,0,0,0,3940
212,0,28,134,3,0,0,0,0,1,3941   213,0,14,135,1,0,0,0,0,0,3941

214,0,28,130,3,0,0,0,0,0,3969   215,0,25,120,1,0,0,0,0,2,3983
216,0,16,95,3,0,0,0,0,1,3997   217,0,20,158,1,0,0,0,0,1,3997
218,0,26,160,3,0,0,0,0,0,4054   219,0,21,115,1,0,0,0,0,1,4054
220,0,22,129,1,0,0,0,0,0,4111   221,0,25,130,1,0,0,0,0,2,4153
222,0,31,120,1,0,0,0,0,2,4167   223,0,35,170,1,0,1,0,0,1,4174
224,0,19,120,1,1,0,0,0,0,4238   225,0,24,116,1,0,0,0,0,1,4593
226,0,45,123,1,0,0,0,0,1,4990

4. Relying on a univariate analysis for classification purposes may not always be wise. The following table provides the intervals in minutes between successive eruptions of the Old Faithful geyser in Yellowstone National Park, and is taken from Härdle [1991].

   a. Is the distribution of intervals actually a mixture of distributions? If so, how many?

   b. Successive intervals appear to be dependent. To characterize this dependence, form a series of pairs consisting of the lengths of the previous and the current interval, (54,74), (74,62) and so on. Once again, address the following questions; Is the distribution of intervals actually a mixture of distributions? If so, how many? (The table should be read one row at a time).

   54 74 62 85 55 88 85 51 85 54 84 78 47 83 52 62 84 52 79 51 47 78
   69 74 83 55 76 78 79 73 77 66 80 74 52 48 80 59 90 80 58 84 58 73
   83 64 53 82 59 75 90 54 80 54 83 71 64 77 81 59 84 48 82 60 92 78
   78 65 73 82 56 79 71 62 76 60 78 76 83 75 82 70 65 73 88 76 80 48
   86 60 90 50 78 63 72 84 75 51 82 62 88 49 83 81 47 84 52 86 81 75
   59 89 79 59 81 50 85 59 87 53 69 77 56 88 81 45 82 55 90 45 83 56
   89 46 82 51 86 53 79 81 60 82 77 76 59 80 49 96 53 77 77 65 81 71 70
   81 93 53 89 45 86 58 78 66 76 63 88 52 93 49 57 77 68 81 81 73 50 85
   74 55 77 83 83 51 78 84 46 83 55 81 57 76 84 77 81 87 77 51 78 60 82
   91 53 78 46 77 84 49 83 71 80 49 75 64 76 53 94 55 76 50 82 54 75 78
   79 78 78 70 79 70 54 86 50 90 54 54 77 79 64 75 47 86 63 85 82 57 82
   67 74 54 83 73 73 88 80 71 83 56 79 78 84 58 83 43 60 75 81 46 90 46
   74

5. Apply Silverman's method to Fisher's Iris data to determine the number of species.

Petal Length

1.4, 1.4, 1.3, 1.5, 1.4, 1.7, 1.4, 1.5, 1.4, 1.5, 1.5, 1.6, 1.4, 1.1, 1.2,
1.5, 1.3, 1.4, 1.7, 1.5, 1.7, 1.5, 1, 1.7, 1.9, 1.6, 1.6, 1.5, 1.4, 1.6,
1.6, 1.5, 1.5, 1.4, 1.5, 1.2, 1.3, 1.4, 1.3, 1.5, 1.3, 1.3, 1.3, 1.6, 1.9,
1.4, 1.6, 1.4, 1.5, 1.4, 4.7, 4.5, 4.9, 4, 4.6, 4.5, 4.7, 3.3, 4.6, 3.9,
3.5, 4.2, 4, 4.7, 3.6, 4.4, 4.5, 4.1, 4.5, 3.9, 4.8, 4, 4.9, 4.7, 4.3,
4.4, 4.8, 5, 4.5, 3.5, 3.8, 3.7, 3.9, 5.1, 4.5, 4.5, 4.7, 4.4, 4.1, 4,
4.4, 4.6, 4, 3.3, 4.2, 4.2, 4.2, 4.3, 3, 4.1, 6, 5.1, 5.9, 5.6, 5.8, 6.6,
4.5, 6.3, 5.8, 6.1, 5.1, 5.3, 5.5, 5, 5.1, 5.3, 5.5, 6.7, 6.9, 5, 5.7,
4.9, 6.7, 4.9, 5.7, 6, 4.8, 4.9, 5.6, 5.8, 6.1, 6.4, 5.6, 5.1, 5.6, 6.1,
5.6, 5.5, 4.8, 5.4, 5.6, 5.1, 5.1, 5.9, 5.7, 5.2, 5, 5.2, 5.4, 5.1

6. Suppose you've collected the prices of a number of stocks over a period of time and have developed a model with which to predict their future behavior. How would you go about validating your model?

CHAPTER 11

Survival Analysis and Reliability

11.0.   Introduction

Survival analysis and reliability are often treated as separate disciplines, each with its own terminology and literature. But both involve the breakdown of items over time, although survival analysis deals with individuals (e.g., AIDS patients) and reliability with things (e.g., light bulbs, disk drives).

Survival analysis/reliability is of particular interest to the statistician. On the one hand, the data form a counting process in which the number of remaining items is a function of time; on the other, the data consists of continuous variables whose values may be censored.

In this chapter, you'll use both approaches to estimate median survival times (median time to failure) and to compare the efficacy of two treatments (two manufacturing processes).

TABLE 11.1. Failure Time in Hours as a Function of Temperature

Temperature	Insulation			Temperature	Insulation		
	I	II	III		I	II	III
200° for 14 days	1176	2520	3528	250° for 1 day	204	300	252
	1512	2856	3528		228	324	300
	1512	3192	3528		252	372	324
	1512	3192			300	372	
	3528	3528			324	444	

11.1. Data Not Censored

Many problems involving survival functions and failure rates can be resolved by methods with which we are already familiar. As one example, Table 11.1, adapted from Nelson [1982], compares the heat resistance of three types of insulators. The appropriate method of analysis is given in Chapter 8 on experimental design: We treat each temperature condition as a separate block and apply any of the statistics in Section 8.2.1.

11.1.1. Savage Scores

Deaths (failures) occur at irregular points in time; there might be $d_1 =$ one death on the first day, none on the second through fifth, and then $d_2 =$ three deaths on the sixth. If the data is not censored and we can assume the death/failure rate is constant at all ages, the optimal statistic would make use of the expected order statistics from the exponential distribution adjusted for ties, also known as *Savage scores*:

$$w_j = \frac{1}{d_j}[\sum_{m=h_j}^{v_j} \sum_{k=1}^{m}(N - k - 1)^{-1}] - 1 \text{ for } j = 1, \ldots c$$

where d_j is the number of failures or deaths that occur on the jth day failures occur, $v_j = d_1 + d_2 + \cdots d_j$, and $h_j = v_{j-1} + 1$.

Table 11.2 records hematologic toxicity measured as the days on which the white blood cell count (WBC) fell below 500 for patients on one of five different drug regimens. Each regimen receives a score that is the sum of the Savage scores of patients on that regimen. The test statistic is F_2, defined in Equation 8.3, but using Savage scores in place of the original observations.

To execute the analysis with the aid of StatXact-3, we select from the pull-down menus **Statistics, K Independent Samples, Savage Scores**, and **Exact test method**. The null hypothesis is that treatments have equal toxicity.

Savage scores can also be used when the data is categorical rather than continuous. We first considered the data in Table 11.3 in Chapter 7, where we applied permutation methods using as the test statistic an ANOVA with arbitrary scores. If we can assume that the underlying processes are exponential, the Savage scores provide the optimal test statistic. To execute the analysis with the aid of StatXact-3, we select in turn **Statistics, Singly Ordered R x C Table, Savage Scores**, and

TABLE 11.2. Hematologic Toxicity

Drug Regimen	Days with WBC< 500
1	0 1 8 10 0
2	0 0 3 3 8
3	5 6 7 14 14
4	1 1 6 7 7 8 8 10
5	7 10 11 12 12 13

TABLE 11.3. Response to Chemotherapy.

	None	Partial	Complete
CTX	2	0	0
CCNU	1	1	0
MTX	3	0	0
CTX+CCNU	2	2	0
CTX+CCNU+MTX	1	1	4

Exact test method. Our hypothesis is that the five rows are drawn from identical exponential populations. The results reveal a statistically significant difference.

```
Datafile: C:\SX3WIN\EXAMPLES\TUMOR.CY3

SAVAGE TEST [That the 5 rows are identically distributed]

Statistic based on the observed data :
  The Observed Statistic = 9.337

Monte Carlo estimate of p-value :
  Pr { Statistic .GE. 9.337 } = 0.0283
  99.00% Confidence Interval = ( 0.0240, 0.0326 )
```

11.2.   Censored Data

The vast majority of survival and reliability studies cannot be followed to completion, particularly in situations where economic and social factors dictate the need for an early decision. As a result, one or more observations in every trial are certain to be *censored*.

Two types of censoring occur. *Type I* censoring is practiced in reliability (mean time to failure) and animal survival studies, where all items/subjects are treated at the same point of time and the experiment is terminated after a fixed interval. *Type II* censoring occurs in clinical trials in which patients enter the study on a continuous basis and the experiment is terminated at some fixed point in time, so that the duration any particular patient remains in the study is a random variable. A study may also be *truncated*. For example, in an epidemiological study of heart-valve patients, we limit our investigations to those who are still alive at some fixed

---

Data Not Censored

Apply traditional tests as in Chapter 8.
If there are many ties and you can assume the death/failure rate is constant at all ages, transform initially using Savage scores, and then apply traditional tests.

point in time. If the study is brought to an end a year from now, it may be both truncated and censored.

11.3.   Type I Censoring

When observations are censored, the most powerful test typically depends on the alternative. Thus, it is not possible to obtain a uniformly most powerful test. If Type I censoring was used it may still be possible to obtain a permutation test that is close to the most powerful test, that is, "almost most powerful," regardless of the underlying parameter values [Good, 1989, 1991, 1992]. The drawback of this globally, almost most powerful test (GAMP), is that it yields a set of bounds on the significance level rather a single value. The good news is that as the sample size grows larger, the bounds grow closer together and, for very large samples, yield an UMP unbiased test [Good, 1992].

Suppose we continue taking observations until time C and then stop. Let S_U and N_C denote the sum of the uncensored observations and the number of censored observations in the treatment sample, respectively. To test the null hypothesis that the two treatments are equivalent against the alternative that the treatment extends the life span, we assign a rearrangement π to the rejection region if $S_U(\pi) - S_U + (N_C(\pi) - N_C)C \geq 0$. We assign π to the acceptance region if $S_U(\pi) - S_U + (N_C(\pi) - N_C)C < 0$. We assign π to a region of indifference otherwise. The fraction of permutations in the rejection region provides us with a lower bound on the significance level, the fraction in the acceptance region with an upper bound.

Table 11.4 records a subset of the intervals between failures for an online real-time transaction system at the beginning and midway through the debugging phase. Management is skeptical as to whether any real progress is being made, and there are rumors the project might be canceled.

One possibility for an analysis is to convert the data to ranks, giving all the censored observations the same value, and, after correcting for ties, use permutation methods to assess significance. The sum of the ranks in the initial sample is $1 + 3 + 2 + 9 + 5 = 20$, which could occur 5.5% of the time by chance, a result that might be thought of as marginally significant.

Applying the globally almost most powerful test, $S_U = 11,920$ and $N_C = 2$ for the initial sample; the sum $S_U + 10,000 N_C$ is equaled or exceeded in 40 of the $\binom{10}{5} = 252$ rearrangements, providing a lower bound on the significance level of 16% and we conclude there has been no significant progress in removing flaws from the software.

TABLE 11.4. Intervals between Failures (in Seconds) for an On-Line Transaction System.

Initial	880,	3,430,	2860,	10,000+,	4,750
Midway	3,820,	10,000+,	7,170,	10,000+,	4,800

Type I Censoring

H : mean/medians of groups are the same.
K : mean/median of treatment group is larger.

Assumptions

1. Under the hypothesis, observations are exchangeable.
2. Observations cannot exceed C in magnitude.

Procedure

1. Compute S_U, the sum of the uncensored observations in the treatment sample.
2. Record N_C, the number of censored observations in the treatment sample.
3. Compute sum $S_U + N_C$ for all possible rearrangements of the combined sample.

Draw a conclusion: Reject the hypothesis if $S_U + N_C$ for the treatment sample is an extreme value.

11.4. Type II Censoring

For clinical trials that use Type II censoring, we may use any of several alternate resampling methods including bootstrap confidence intervals, logistic regression, and a variety of permutation tests.

11.4.1. Bootstrap

Suppose we have n pairs of observations $\{(t_1, \delta_1), \cdots, (t_n, \delta_n)\}$, as in Table 11.5 where t_i is an individual's age at death if the observation is uncensored and his age at the end of the study if the data is censored, while

$$\delta_i = \begin{cases} 1 & \text{if } t_i \text{ is uncensored.} \\ 0 & \text{if } t_i \text{ is censored.} \end{cases}$$

Suppose the times $\{t_i\}$ are already ordered so that $t_1 < t_2 < \cdots < t_n$. One estimate of the survival function due to Kaplan and Meir [1958] is

$$S(t) = \prod_{j=1}^{k_t} \left(\frac{n-i}{n-i+1} \right)^{d_i}$$

where t lies in the interval between the kth and the kth+1st event.[1] The estimated median survival time t_{50} is the value of t for which $S(t) = .5$. We may use the bootstrap to obtain a confidence interval for this estimate as follows.

---

[1] $\prod_{i=1}^{n} x_i = x_1.x_2\cdots x_n.$

TABLE 11.5. Time in Study Following Treatment for Cancer.

Time	Alive	Age at Start	Time	Alive	Age at Start
1		61	8		52
1		65	8		49
2		59	11		50
3		52	11		55
4		56	12		49
4		67	12		62
5		63	15		51
5		58	17		49
8		56	22		57
8	Y	58	23		52

Consider each pair $y_i = (t_i, \delta_i)$ as an independent random sample from a bivariate distribution F. Draw a random sample with replacement from these pairs and use it to estimate t_{50}. Repeating this bootstrapping process 400 times, we obtain a confidence interval for t_{50}.

The preceding analysis assumes that the death rate is the same for all individuals. The bootstrap approach can also be used to obtain confidence intervals for *hazard rates* $h(t|X)$ that vary from person to person. If the probability $h(t|X)$ for an individual with the vector of characteristics X (race, sex, age, blood chemistries) that an event (remission, failure, death) occurs for the first time after an interval t has the proportional hazards form suggested by Cox [1972]

$$h(t|X) = h_o(t)e^{\beta X}$$

the regression parameter β can be estimated independently of the decay function $h(t)$. One commonly used estimate is the "partial likelihood"

$$\hat{\beta} = \prod_{i \in D} \frac{e^{\beta x_i}}{\sum_{k \in R_i} e^{\beta x_k}}$$

where D is the set of indices of failure times, R_i is the set of indices of those at risk at time t_i. To obtain a bootstrap estimate of the standard error of $\hat{\beta}$ or a confidence interval for $\hat{\beta}$, we sample with replacement from the triples $\{(t_1, x_1, \delta_1), \cdots, (t_n, x_n, \delta_n)\}$ and then apply any of the methods given in Chapter 5.

The bootstrap can be applied to a multivariate counting process that is both left-truncated and right-censored [Andersen et al., 1993]. The survival function of this process is extremely complex, requiring page after page of formulas, but the bootstrap confidence intervals for the function require no more effort than in either of our previous examples.

11.4.2. Comparing Two Survival Curves

Forsythe and Frey [1970]) propose a permutation test for comparing two survival curves based on the maximum difference in percentage between the two curves.

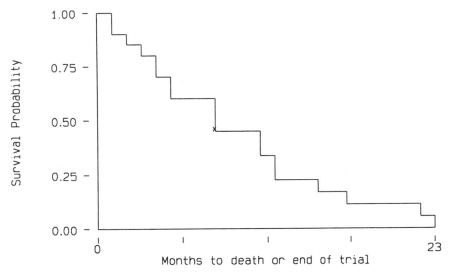

FIGURE 11.1. Kaplan-Meir Survival Curve for Treated Patients.

TABLE 11.6. Survival Times in Months of All Cases of Carcinoma of the Tongue Treated at UCLA between 1955 and 1964.

Males	0–1	1–2	2–3	3–4	4–5	5–6	6–7	7–8	8–9	9–10
Alive	4	7	5	1	1	1	0	0	1	0
Dead	23	17	7	2	2	0	0	1	0	0
Females										
Alive	1	6	3	3	3	3	1	1	0	0
Dead	21	4	1	1	0	0	0	1	0	0

The first step in computing S for the data in Table 11.6 is to compute the cumulative proportion surviving for each sex for each year, as shown here:

Surviving to End of Year	Males	Females
1	.671	.567
2	.396	.473
3	.246	.442
4	.188	.404
5	.120	.404
6	.120	.404
7	.120	.404
8	.060	.135
9	.060	.135
10	.060	.135

The maximum difference occurs in year 5 and is 0.284.

Next, we rearrange the data, preserving the total number of alive and dead for each survival time and the total number of each sex for each survival time. Table 11.6 displays one such possible rearrangement.

TABLE 11.7. Possible Rearrangement of Labels for UCLA Survival Data in Years.

Males	0–1	1–2	2–3	3–4	4–5	5–6	6–7	7–8	8–9	9–10
Alive	5	7	5	2	2	1	0	0	1	0
Dead	22	17	7	1	1	0	0	1	0	0
Females										
Alive	0	6	3	2	2	3	1	1	0	0
Dead	22	4	1	2	1	0	0	1	0	0

Forsythe and Frey [1970] found that only 4% of a sample of 1000 rearrangements had values of the test statistic greater than that observed in the original sample and concluded that men were less likely than women to respond successfully to treatment.[2]

A comparison like this is appropriate only if the data for the two groups has first been blocked or matched in terms of age, stage, past treatment, and other relevant covariates.

11.4.3.   Hazard Function

Seldom are we willing to assume that the hazard function is independent of age. In general, for a two-sample comparison, our null hypothesis is that the two sets of observations come from the same unknown distribution F, while under the alternative, the hazard functions $\lambda_1(u) = \frac{f_1(u)}{1-F_1(u)}$, $\lambda_2(u) = \frac{f_2(u)}{1-F_2(u)}$ satisfy one of the following alternatives:

$$K_1 \text{ (decreasing hazards): } \lambda_1(u) = \theta(u)\lambda_2(u)$$
$$\text{where } \theta(u) \text{ is a decreasing function of } u$$

or

$$K_2 \text{ (proportional hazards): } \lambda_1(u) = \theta\lambda_2(u).$$

If the data is uncensored or if there is equal censoring in the populations, then one might use the generalized Wilcoxon-Gehan test to detect a decreasing hazards alternative and the log rank test to detect proportional hazards.

Suppose d_i is the number of subjects who die at time t_i. Define the Wilcoxon-Gehan score of these subjects as

$$w_{t_i} = 1 - \frac{2}{d_i} \sum_{i=v_i+1}^{v_{i+1}} \prod_{j=1}^{i} \frac{N - C_j - j + 1}{N - C_j - j + 2}$$

where $v_i = d_1 + d_2 + \cdots d_i$.

---

[2] I used to chew cigars. Can you guess why I gave up this habit?

All C_j subjects who are censored in the time interval (t_i, t_{i+1}) receive the same Wilcoxon-Gehan score

$$w_{t_i+} = 1 - \prod_{j=1}^{i} \frac{N - C_j - j + 1}{N - C_j - j + 2}$$

where N is the total number of subjects in the study. The corresponding log rank scores are

$$w_{t_i} = 1 - \left(\frac{2}{d_i} \sum_{i=v_i+1}^{v_{i+1}} \sum_{j=1}^{i} \frac{1}{N - C_j - j + 1}\right) - 1$$

$$w_{t_i+} = \sum_{j=1}^{i} \frac{1}{N - C_j - j + 1}.$$

The test statistics consist of the sums of the scores for all observations, censored and uncensored.

11.4.4. An Example

The results of a clinical trial on lung cancer patients are recorded in Table 11.7 in which a new treatment is compared with an existing one. Two of the patients are still living at the end of the trial, (a *yes* indicates that the corresponding survival time is censored). The log rank and Wilcoxon-Gehan scores are also included in the table.

Datafile: C:\SX3WIN\EXAMPLES\CANCER.CY3

LOGRANK TEST

Exact Inference:

TABLE 11.8. Lung Cancer Clinical Trial. Survival in Days.

Drug	DAYS	Censored	Logrank	W-G
New	250		−0.19843767	0.06666667
New	476	Yes	0.41822899	0.73333333
New	355		0.41822899	0.46666667
New	200		−0.46629482	−0.20000000
New	355	Yes	1.41822899	0.73333333
Old	191		−0.57740593	−0.33333333
Old	563		1.41822899	0.73333333
Old	242		−0.34129482	−0.06666667
Old	285		0.16822899	0.33333333
Old	16		−0.80648518	−0.66666667
Old	16		−0.80648518	−0.66666667
Old	16		−0.80648518	−0.66666667
Old	257		−0.03177101	0.20000000
Old	16		−0.80648518	−0.66666667

```
One-sided p-value:
  Pr { Test Statistic .GE. Observed } = 0.0529
Two-sided p-value:
  Pr { | Test Statistic  - Mean | .GE. | Observed - Mean |
    = 0.0974
```

WILCOXON-GEHAN TEST

```
Exact Inference:
  One-sided p-value:
    Pr { Test Statistic .GE. Observed } = 0.0380
  Two-sided p-value:
    Pr { | Test Statistic - Mean | .GE. | Observed - Mean |
      = 0.0729
```

As you can see, a difference in scoring system can alter the interpretation of the result. Determine which hazard function is appropriate before you do your analysis.

11.5.   Summary and Further Readings

In this chapter, we applied the methods of previous chapters to the analysis of survival and reliability data that can be expressed in terms of either survival times or surviving numbers. We distinguished between censored and uncensored data and between Type I and Type II censoring. We introduced the GAMP test for Type I censored data, and the Kaplan-Meir bootstrap, Forsythe-Frey, logrank, and Wilcoxon-Gehan permutation tests for Type II.

The application of permutation methods to censored data was first suggested by Kalbfleisch and Prentice [1980], who sampled from the permutation distribution of censored data to obtain estimates in a process akin to bootstrapping. Hazard functions and partial likelihood were introduced by Cox [1972, 75]. See also Kleinbaum, Kupper and Chambless [1982].

Permutation test are applied to censored data by Macuson and Nordbrock [1981], Koziol et al. [1981], Jennrich [1983, 84], Schemper [1984], Hiriji, Mehta, and Patel [1987]. Conditional rank tests for randomly censored survival data are described by Andersen et al. [1982] and Janssen [1991]. Bootstrap applications include those of Efron [1981], Akritas [1986], and Altman and Andersen [1989]. Gordon and Olshen [1985] use a tree-structured analysis.

The counting process approach to survival analysis and generalized transition models is due to Aalen [1978] with excellent reviews in Andersen et al. [1992] and Flemington and Harrington [1991].

11.6.   Exercises

1. Estimate the median survival time for the data of Figure 11.1. Provide a bootstrap confidence interval for this estimate.
2. Graph the survival distribution for the data of Table 11.5. Estimate the 80th percentile for males. Provide a confidence interval for this estimate.
3. **a.** Section 11.2 makes reference to a hypothetical study of heart-valve patients in which those still living at the start of the investigation were followed for an additional year. Is this study an example of Type I or Type II censoring?
   b. Most instruments and assays, though linear in the center of their range, tend to be nonlinear and unreliable when used to measure extremely small or extremely large values. Would the readings from such instruments undergo Type I or Type II censoring?
4. RFM male mice exposed to 300 rads X-radiation at 5 to 6 weeks of age were followed for a year while living either in a conventional laboratory or a germ-free environment [Hoel and Walburg, 1972]. At the end of the year, individual ages at death were as follows:

   Conventional environment 159, 189, 191, 198, 200, 207, 220, 235, 245, 250, 256, 261, 256, 261, 265, 266, 280, 343, 356 plus five survivors.
   Germ-free environment 158, 192, 193, 194, 195, 202, 212, 215, 229, 230, 237, 240, 244, 247, 259, 300, 301, 321, 337, plus ten survivors.

   a. What type of censoring was used?
   b. Plot the survival curves for the two groups.
   c. Estimate the median survival time for each group. Provide a confidence interval for your estimate.
   d. Is there a significant difference in the survival times of the two groups?
   e. Is age at death the appropriate measure of survival time?
   f. The data presented are solely for mice who died of thymic lymphoma. Would it be better to include the data from all irradiated animals?

5. The data in Table 11.4 is actually that of a much larger study in which several cancer treatments are compared. Following is the data for a second drug. Which drug is more effective? If several different methods are applicable, compare them (e.g., bootstrap confidence intervals versus log rank).

Time	Alive	Age at Start	Time	Alive	Age at Start
6		67	15	Y	50
6	Y	65	16		67
7		58	19	Y	50
9	Y	56	20	Y	55
10	Y	49	22		58
11	Y	61	23		47
13		62	32	Y	52

6. Table 11.1 is somewhat suspicious. Although not noted by Nelson [1982], the repeated occurrence of 3,528 suggests that the data is actually Type I censored

with the last observation being made at 3,528 hours. Analyze the experiment as if this were actually the case.

7. Time in service need not be the sole measure of reliability. For example, Grubbs [1971] records the mileage between failures of 14 personnel carriers:

162, 200, 271, 302, 393, 508, 539, 629, 706, 777, 884, 1,008, 1,101, 1,182, 1,463, 1,603, 1,984, 2,355, 2,880.

   a. What is your estimate of the median mileage between failures?
   b. Provide a confidence interval for your estimate.
   c. Provide a confidence interval on the reliability at 200 miles.
   d. Test the hypothesis that the mean mileage between failures is 800.

8. Ten women underwent radical mastectomies: 5 of the 10 chosen at random were given an experimental drug, while the remaining 5 were left untreated. The 10 were followed for a two-year period, and the tumor-free interval recorded. Is the drug effective?

| Drug Group: | 23, | 16*, | 18*, | 20* | |
| Placebo: | 15, | 18, | 19, | 19, | 20 |

Asterisks denote patients with no detectable tumor at the conclusion of the study. (*Hint*: Is this Type I or Type II censoring?)

9. How would you go about testing whether a failure rate actually was constant? (*Hint*, see Chapter 10).

Which Statistic Should I Use?

This chapter provides you with an expert system for use in choosing an appropriate estimation or testing technique. Your expert system comes to you in two versions— a professional's handbook with detailed explanations of the choices and a quick-reference version at the end of the chapter.

12.1.   Parametric versus Nonparametric

One should use a parametric test:

- When you have a large number of observations (≥ 40) in each category; use Student's t, or the F (see Chapter 4).
- When you have a very small number of observations (≤ 5) if the assumptions underlying the corresponding parametric test may be relied on (see Section 12.2). [1]

Apart from these two exceptions, the resampling approach has several advantages over the parametric:

- The permutation test is exact under relatively nonstringent conditions. In the one-sample problem (which includes regression), observations must have symmetric distributions; in the 2- and k-sample problem, observations must be exchangeable among the samples.

---

[1] The rationale is that we may not have enough rearrangements to resample successfully. For example, if we have only three observations, the sample space for the permutation test is limited to 2^3 or 8 rearrangements. As a result, we must randomize on the boundary except for significance levels that are multiples of $1/8$.

- The permutation test provides protection against deviations from parametric assumptions, yet is usually as powerful as the corresponding unbiased parametric test, even for small samples.

With two binomial or two Poisson populations, the most powerful unbiased permutation test and the most powerful parametric test coincide. With two normal populations, the most powerful unbiased permutation test and the most powerful unbiased parametric test produce identical results for very large samples.

- Using a resampling method means you are no longer dependent on the availability of tables but you can choose the statistic that is best adapted to your problem and to the alternatives of interest.

Consider a permutation test before you turn to a bootstrap. The bootstrap is not exact except for quite large samples and, often, is not very powerful. But the bootstrap can sometimes be applied when the permutation test fails: Examples include unbalanced designs and tests of ratios and proportions.

12.2.   But Is It a Normal Distribution?

To perform a parametric test, we must assume that the observations come from a probability distribution that has a specific parametric form such as the binomial, Poisson, or Gaussian (normal) described in Section 2.2.

While there exist various formal statistical techniques for verifying whether a set of observations does or does not have a Poisson or normal distribution, the following informal guidelines are of value in practice:

- An observation has the Poisson distribution if it is the cumulative result of a large number of opportunities, each of which has only a small chance of occurring. For example, if we seed a small number of cells into a petri dish that is divided into a large number of squares, the distribution of cells per square follows the Poisson.
- An observation has the Gaussian or normal distribution if it is the sum of a large number of factors, each of which makes a very small contribution to the total. This explains why the mean of a large number N of observations, $\overline{X} = \frac{1}{N} \sum_{i=1}^{N} X_i$ will be normally distributed even if the individual observations come from non-Gaussian distributions.

In many applications in economics and pharmacology, where changes are best expressed in percentages, a variable may be the product of a large number of variables, each of which makes only a very small contribution to the total. Such a variable has the log normal distribution and, as $\log(X_1 X_2 \cdots X_n) = \sum \log(X_i)$, its logarithm has a normal distribution.

12.3. Which Hypothesis?

Recall that with a permutation test, we:

1. Choose a test statistic $S(X)$.
2. Compute S for the original set of observations. (We may first transform to ranks to minimize the effects of a few extreme observations.)
3. Obtain the resampling distribution of S by repeatedly rearranging the data. With two or more samples, we combine all the observations into a single large sample before we rearrange them.
4. Accept or reject the null hypothesis according to whether S for the original observations is smaller or larger than the upper α-percentage point of the resampling distribution.

To obtain a nonparametric bootstrap, we:

1. Choose a test statistic $S(X)$.
2. Compute S for the original set of observations.
3. Obtain the bootstrap distribution of S by repeatedly resampling from the observations. We need not combine the samples but may resample separately from each sample. We resample with replacement. (Afterward we may smooth, correct for bias, accelerate, or iterate to obtain a more accurate resampling distribution.)
4. Accept or reject the null hypothesis according to whether S for the original observations is smaller or larger than the upper α-percentage point of the (revised) resampling distribution.

To obtain a parametric test, we:

1. Choose a test statistic S whose distribution F may be computed and tabulated independent of the observations.
2. Compute S for the observations X. (We may use an initial transformation to make the various samples more alike in dispersion.)
3. Accept or reject the null hypothesis according to whether $S(X)$ is smaller or larger than the upper α-percentage point of F.

In almost all the examples we've considered so far, the statistics S used in the parametric, permutation, and bootstrap methods were similar: the difference lay in the methods by which critical values were determined. When comparing the variances of two populations, the nature of the hypothesis, the form of the test statistic, and the method all depend on the approach we use.

Parametric: To test the hypothesis H_1 that both populations are normal with mean 0 and variance σ^2 against the alternative K_1 that the second population, while still normal with mean 0, has variance $\sigma_1^2 > \sigma^2$, the best test is a parametric one based on the distribution of the ratio of the sample variances.

Permutation: To test the hypothesis H_2 that both populations have the same distribution F against the alternative K_2 that one population has the distribution G with scale parameter σ such that $G[x/\sigma] = F[x]$, the best test is a permutation test based on the resampling distribution of Good's test statistic described in Section 3.4.

Bootstrap: To test the hypothesis H_3 that both populations have the same scale parameter σ against the alternative K_3 that one population has the scale parameter $\sigma_1 > \sigma$, the best test is based on the bootstrap estimate of the ratio of the two parameters.

Density estimation: To test whether the dispersions of the two distributions are the same (the variances need not even exist), the best test is based on the permutation distribution of Aly's test statistic considered in Section 3.4.

12.4.   A Guide to Selection

The initial division of this guide is into three groupings: categorical data, discrete data, and continuous data.

12.4.1.   Data in Categories

Examples include men versus women, white versus black versus Hispanic versus other, and improved versus no change versus worse.

Only a single factor is involved. You are testing the goodness of fit of a specific model. See Section 10.6.

Only two factors are involved. For example, sex versus political party. Each factor is at exactly two levels. There is a single table. Use Fisher's exact test (See Section 7.1). There are several 2×2 tables. See Section 7.2.1.

One factor is at three or more levels. This factor is not ordered (e.g. race). You want a test that provides protection against a broad variety of alternatives. Use the permutation distribution of the chi-square statistic (Section 7.4). You wish to test against the alternative of a cause-effect dependence. Use the Freeman and Halton test (Section 7.3.1).

This factor can be ordered. Assign scores to this factor based on your best understanding of its effects on the second variable (Section 7.5.1). Use Pitman correlation (Section 7.5.1).

Both factors are at three or more levels. Neither factor can be ordered. One factor might be caused or affected by the other. Use permutation distribution of Kendall's τ or Cochran's Q (Section 7.4). Cause and effect relationship not suspected. Use permutation distribution of χ^2 statistic (Section 7.4).

One factor can be ordered. Assign scores to this factor based on your best understanding of its effects on the second variable (Section 7.5.1). Use the ANOVA with arbitrary scores (Section 7.5.2.1).

Both factors can be ordered. Use Mantel's U (Section 7.5.2.2). A third covariate factor is present. Use the method of Bross [1964].

12.4.2.   Discrete Observations

The data are discrete, taking values in the ordered set 0, 1, 2, ... Each sample consists of a fixed number of independent identically distributed observations, which can be either 0 or 1. (A set of trials, each of which may result in a success or a failure is an example.)

One or two samples. Use the parametric test for the binomial. See for example, Lehmann [1986, p. 81, 154]. A confidence interval for the binomial parameter may be obtained using the method of Section 5.2.1.

More than two samples, but only one factor. Analyze as indicated earlier categorical data.

More than one factor. Transform the data to equalize variances. For each factor combination, take the arcsin of the square root of the proportion of observations that take the value 1. Analyze as in Section 12.4.3.

Each sample consists of a set of independent identically distributed Poisson observations.

One or two samples. Use the parametric test for the Poisson. In the two-sample case, the UMPU test uses the binomial distribution, Section 4.2.2.

More than two samples. Transform the data to equalize the variances by taking the square root of each observation. Analyze as in Section 12.4.3.

Each sample consists of a set of exchangeable observations whose distribution is unknown.

Single sample. Data may be assumed to come from a symmetric distribution. Use permutation test for a location parameter, Section 3.7.2. The data may not be assumed to come from a symmetric distribution. Use bootstrap described in Section 3.7.1. If you have only a few subjects, consider using a multivariate approach (see Chapter 9).

Two samples or more. Use one of the permutation tests designed for data with continuous distributions in Section 12.4.3. Treat tied observations as distinct observations when you form rearrangements. Be cautious in interpreting a negative finding; the significance level may be large simply because the test statistic can take on only a few distinct values.

12.4.3.   Continuous Data

How precise do measurements have to be so we may categorize them as "continuous" rather than discrete? Should they be accurate to two decimal places as in 1.02? Or four as in 1.0203? To apply statistical procedures for continuous variables, the observations need only be precise enough that there are no or only a very few ties.

If you recognize the data has the normal distribution, a parametric test like Student's t or the F-ratio may be applicable. But you can protect yourself against deviations from normality by making use of the resampling distribution of these statistics.

SINGLE SAMPLE

You want to estimate the population median. Use the sample median; for a confidence interval, see Section 5.2.2.

You want to estimate the mean, standard deviation, correlation coefficient, or some other population parameter. Use the corresponding plug-in estimate and apply the methods of Chapter 5 to obtain a bootstrap confidence interval.

You want to test that the location parameter has a specific value and feel safe in assuming the underlying distribution is symmetric about the location parameterUse the permutation test described in Section 3.7.2.

If the distribution is not symmetric but has a known parametric form, apply the corresponding parametric test. If the distribution is not symmetric and does not have a known parametric form, consider applying an initial transformation that will symmetrize the data. For example, take the logarithm of data that undergoes percentage changes. Be aware that such a transformation affects the form of the loss function.

Otherwise, bootstrap (see Section 3.7.1 and the refinements described in Sections 5.4 and 5.7–5.9)

You want to test that the scale parameter has a specific value. First, divide each observation by the hypothesized value of the scaleparameter. Then, apply one of the procedures noted earlier for testing a location parameter. You want to assess the effects of covariates, bootstrap (see Section 5.4).

TWO SAMPLES

You want to test whether the scale parameters are equal. You know the means/medians of the two populations or you know they are equal. Use the permutation-distribution of the ratio of the sample variances. You have no information about the means/medians. Use Aly's test (Section 3.4).

You want to test whether the location parameters of the two populations are equal. If changes are proportional rather than additive, use the logarithms of the observations. If you suspect outliers, use ranks. If the data have been Type I censored, see Section 11.3. If the data have been Type II censored, see Section 11.4. Each sample consists of measures taken on different subjects. Use the two-sample comparison described in Section 3.1. For a confidence interval for the difference in location parameters, see Section 4.2.2. For a confidence interval for the ratio of two parameters, see Section 5.4.

Two observations were made on each subject; these observations are to be compared. Use the matched-pair comparison described in Sections 3.7.2 and 3.9. To obtain a confidence interval for the difference, see Section 5.2.2.

MORE THAN TWO SAMPLES

If changes are proportional rather than additive, work with logarithms of the observations. A single factor distinguishes the various samples. You can't take advantage of other factors to block the samples (see Section 8.1.1).

Factor levels not ordered. Use permutation distribution of an F-ratio (see Section 8.2.1).

Factor levels are ordered. Use Pitman's correlation (Section 3.5).

You can take advantage of other factors to block the samples. Rerandomize on a block-by-block basis, then apply one of the techniques described in Section 8.2.1.

Multiple factors are involved. One of the factors consists of repeated measurements made over time; see Good [1994, p. 75].

All observations are exchangeable. The experimental design is balanced. Use one of the permutation techniques described in Section 8.2.

Experimental design not balanced. Some factors will be confounded. Use bootstrap (Section 8.5).

12.5. Quick Key

12.5.1. Categorical Data

Single factor, one row
    Goodness of fit (10.4)
    Two factors, two rows
        Two levels (columns):
            Single table: Fisher's exact test (7.1)
            Several 2×2 tables (7.2.1)
        Three or more levels
            Not ordered
                Permutation distribution of χ^2 (7.4)
            Ordered.
                Assign scores (7.5.1)
                Use Pitman correlation (7.5.1)
        Both factors at three or more levels
            Neither factor ordered
                One factor might be caused or affected by the other
                    Use Kendall's τ or Cochran's Q (7.4)
                Cause and effect relationship not suspected
                    Use χ-square statistic (7.4)
            One factor can be ordered
                Assign scores (7.5.1)
                Use ANOVA with arbitrary scores (7.5.2.1)
            Both factors can be ordered
                Use Mantel's U (7.5.2.2)
    With third covariate factor, see Bross [1964].

12.5.2. Discrete Observations

Binomial data
    One or two samples
        See Lehmann [1986, p. 81, 154]
    More than two samples, but only one factor
        Goodness of fit (10.4)

More than one factor
   Transform using arcsin of the square root; then analyze as in Section 12.4.3
Poisson data
   One or two samples
      See Section 4.2
   More than two samples
      Transform by taking the square root, then analyze as in Section 12.4.3
Other exchangeable discrete observations
   Single sample
      Symmetric distribution; permute (3.7.2)
      Not symmetric; transform then bootstrap (3.7.1)
   Two or more samples See Section 12.4.3

12.5.3. Continuous Data

SINGLE SAMPLE

Estimate parameters
      Sections 5.1, 5.3, and 5.4
   Test location parameter
      Distribution symmetric (Section 3.7.2)
      Distribution not symmetric, has known parametric form; apply corresponding parametric test
      Distribution does not have known parametric form applying symmetrizing transformation; else bootstrap (Sections 3.7.1 and 5.7–5.9)
   Test scale parameter
      Scale, then test as above

TWO SAMPLES

Test whether scale parameters equal means/medians of two populations known or known to be equal. Use permutation-distribution of ratio of sample variances.
   Else use Aly's test (Section 3.4).
   Test whether location parameters equal changes proportional rather than additive, use logarithms. Suspect outliers, use ranks. Type I censored (Section 11.3); Type II censored (Section 11.4).
   Different subjects (Section 3.1).
   Matched pairs (Sections 3.7.2 and 3.9).

MORE THAN TWO SAMPLES

Changes proportional rather than additive, use logarithms.
   Single factor
      Try to block the samples (see Section 8.1.1).
      Factor levels not ordered (Section 8.2.1).
      Factor levels ordered; Pitman's correlation (Section 3.5).
   Multiple factors

Resample on a block-by-block basis (Section 8.2.1).
Repeated measurements; see Good [1994, p.75].
  Observations exchangeable; experimental design balanced (Section 8.2).
  Experimental design not balanced (Section 8.5).

12.6.   Classification

Determine the number of categories:
  Bump-hunting, see Section 10.2.1.
  Block clustering, see Section 10.3.
  Develop an assignment rule:
  CART, see Section 10.4.
  Other methods, see Section 10.5.
  Validate or cross-validate your results; see Sections 10.6 and 10.7.

APPENDIX 1

Program Your Own Resampling Statistics

Whether you program from scratch in BASIC, C, or FORTRAN or use a macro language like that provided by Stata, S-Plus, or SAS, you'll follow the same four basic steps:

1. Read in the observations.
2. Compute the test statistic for these observations.
3. Resample repeatedly.
4. Compare the test statistic with its resampling distribution.

Begin by reading in the observations and, even if they form an array, store them in the form of a linear vector for ease in resampling. At the same time, read and store the sample sizes so the samples can be re-created from the vector. For example, if we have two samples 1, 2, 3, 4 and 1, 3, 5, 6, 7, we would store the observations in one vector $X = (1\ 2\ 3\ 4\ 1\ 3\ 5\ 6\ 7)$ and the sample sizes in another (4 5) so as to be able to mark where the first sample ends and the next begins.

Compute the value of the test statistic for the data you just read in. If you are deriving a permutation statistic, several shortcuts can help you here. Consider the well-known t-statistic for comparing the means of two samples:

$$t = \frac{\overline{X} - \overline{Y}}{\sqrt{S_x^2 + S_y^2}}.$$

The denominator is the same for all permutations, so don't waste time computing it, instead use the statistic

$$t' = \overline{X} - \overline{Y}.$$

A little algebra shows that

$$t' = \Sigma X_i \left(\frac{1}{n_x} + \frac{1}{n_y} \right) - (\Sigma X_i + \Sigma Y_i) \frac{1}{n_y}.$$

Four Steps to a Resampling Statistic

1. Get the data.
2. Compute the statistic S_o for the original observations.
3. Loop:

   a. Resample.
   b. Compute the statistic S for the Resample.
   c. Determine whether $S > S_o$.

4. State the p-value.

But the sum $(\Sigma X_i + \Sigma Y_i)$ of all the observations is the same for all permutations, as are the sample sizes n_x and n_y; eliminating these terms (why do extra work if you don't need to?) leaves

$$t'' = \Sigma X_i$$

the sum of the observations in the first sample, a sum that is readily computed.

Whatever you choose as your test statistic, eliminate terms that are the same for all permutations before you begin.

How we resample will depend on whether we use the bootstrap or a permutation test. For the bootstrap, we repeatedly sample without replacement from the original sample. Some examples of program code are given in a sidebar. More extensive examples entailing corrections for bias and acceleration are given in Appendix 3.

IMPORTANCE SAMPLING

We may sample so as to give equal probability to each of the observations or we may use importance sampling in which we give greater weight to some of the observations and less weight to others.

Computing a Bootstrap Estimate

C++ Code:

```
#include <stdlib.h>
get_data();  //put the variable of interest in the first n
             //elements of the array X[].
randomize(); //initializes random number generator
for (i=0; i<100; i++){
        for (j=0; j<n; j++) Y[j]=X[random(n)];
        Z[i]=compute_statistic(Y);
        //compute the statistic for the array Y and store
        //it in Z compute_stats(Z);
```

Computing a Bootstrap Estimate

GAUSS:

```
n = rows(Y);
U = rnds(n,1, integer seed));
I = trunc(n*U + ones(n,1,1));
    Ystar = Y[I,.];
```

(This routine only generates a single resample.)

SAS Code:

```
PROC IML
  n = nrow(Y);
  U = ranuni(J(N,1, integer seed));
  I = int(n*U + J(n,1,1));
      Ystar = Y(|I,|);
```

(This routine only generates a single resample. A set of comprehensive algorithms in the form of SAS macros is provided in Appendix 3.)

S-Plus Code:

```
''bootstrap'' <- function(x, nboot, statistic,. . .){
data <- matrix(sample(x,size=length(x)*nboot, replace=T),
                          nrow=nboot)
return(apply(data,1,statistic,. . .))
}
# where statistic is an S+ function which calculates the
# statistic of interest.
```

Stata Code:

```
. use data, clear
. bstrap median, args(height), reps(100)

  where median has been predefined as follows:
program define median
      if '''1''' == ''?'' {
              global S_1 ''median''
              exit
              }
summarize '2',detail
post '1' _result(10)
end
```

The idea behind importance sampling is to reduce the variance of the estimate and, thus, the number of resamples necessary to achieve a given level of confidence.

Suppose we are testing the hypothesis that the population mean is 0 and our original sample contains the observations -2, -1, -0.5, 0, 2, 3, 3.5, 4, 7, and 8. Resamples containing 7 and 8 are much more likely to have large means, so instead of drawing bootstrap samples such that every observation has the same probability $1/10$ of being included, we'll weight the sample so that larger values are more likely to be drawn, selecting -2 with probability $1/55$, -1 with probability $2/55$, and so forth, with the probability of selecting 8 being $9/55$'s.[1] Let $I(S* > 0) = 1$ if the mean of the bootstrap sample is greater than 0 and 0 otherwise. The standard estimate of the proportion of bootstrap samples greater than zero is given by

$$\frac{1}{B} \sum_{b=1}^{B} I(S_b* > 0).$$

Because since we have selected the elements of our bootstrap samples with unequal probability, we have to use a weighted estimate of the proportion

$$\frac{1}{B} \sum_{b=1}^{B} I(S_b* > 0) \frac{\prod_{i=1}^{10} \pi_i^{t_{ib}}}{.1^{10}}$$

where π_i is the probability of selecting the ith element of the original sample and t_{ib} is the number of times the ith element appears in the bth bootstrap sample. Efron and Tibshirani [1993, p. 355] found a sevenfold reduction in variance using importance sampling with samples of size 10.

Importance sampling for bootstrap tail probabilities is discussed in Johns [1988], Hinkley and Shi [1989], and Do and Hall [1991]. Mehta, Patel, and Senchaudhuri [1988] considered importance sampling for permutation tests.

SAMPLING WITH REPLACEMENT

For permutation tests, three alternative approaches to resampling suggest themselves:

- Exhaustive—examine all possible resamples.
- Focus on the tails—examine only samples more extreme than the original.
- Monte Carlo.

The first two of these alternatives are highly-application specific. Zimmerman [1985a,b] provides algorithms for exhaustive enumeration of permutations. The computationally efficient algorithms developed in Mehta and Patel [1983] and Mehta, Patel, and Gray [1985] underlie the StatXact results for categorical and ordered data reported in in Chapter 7. Algorithms for exact inference in logistic regression are provided in Hirlje, Metha, and Patel [1988].

An accompanying sidebar provides the C++ code for estimating permutation test significance levels via a Monte Carlo. Unlike the bootstrap, permutations

---

[1] $1 + 2 + 3 + 4 + 5 + 6 + 7 + 8 + 9 + 10 = 55$.

Estimating Permutation Test Significance Levels

C++ Code for One-Sided Test:

```cpp
int Nmonte; // number of Monte Carlo simulations
float stat0; // value of test statistic for the unpermuted
            // observations
float X[]; // vector containing the data
int n[]; // vector containing the cell/sample sizes
int N; // total number of observation

proc main() {
  int i, s, k, seed, fcnt=0;
  get_data(X,n);
  stat_orig = statistic(X,n); // compute statistic for the
                              // original sample
  srand(seed);

  for (i=0; i < Monte; i++){
    for (s = N; s > n[1]; s--)
      {
      k = choose(s); // choose a random integer from 0 to s-1
      temp = X[k];
      X[k] = X[s-1]; // put the one you chose at end of
                     // vector
      X[s-1] = temp;
      }
    if (stat_orig >= statistic(X,n)) fcnt++;
    // count only if the value of the statistic for the
    // rearranged sample is as or more extreme than its value
    // for the original sample.
    }
  cout << ``alpha = '' << fcnt/NMonte;
}

void get_data(float X, int n){
// This user-written procedure gets all the data and packs
// it into a single long linear vector X. The vector n is
// packed with the sample sizes.
}

float statistic(float X, int n){
// This user-written procedure computes and returns the
// test statistic
{
```

Computing a Single Random Rearrangement

With SAS:

```
PROC IML
n = nrow (Y);
U = randuni (J(n,1,seed));
I = rank (U);
Ystar = Y(|I,|);
```

With GAUSS

```
n = nrows (Y);
U = rnds (n,1,seed);
I = rankindx (U,1);
Ystar = Y[|I,|];
```

---

require that we select elements with replacement. The trick (see sidebar) is to stick each selected observation at the end of the vector, where it will not be selected a second time. Appendix 2 contains several specific applications of this programming approach.

C++, SC, and Stata Code for Permutation Tests

Simple variations of the C++ program provided in Appendix 1 yield many important test statistics.

Estimating Significance Level of the Main Effect of Fertilizer on Crop Yield in a Balanced Design

Set aside space for

Monte	the number of Monte Carlo simulations
S_0	the original value of test statistic
S	test statistic for rearranged data
data	{5, 10, 8, 15, 22, 18, 21, 29, 25, 6, 9, 12, 25, 32, 40, 55, 60, 48};
$n = 3$	number of observations
in each category blocks $= 2$	number of blocks
levels $= 3$	number of levels of factor

Main program:

```
GET_DATA
   put all the observations into a single linear vector
COMPUTE S_0 for the original observations
Repeat Monte times:
  for each block
     REARRANGE the data in the block
  COMPUTE S
  Compare S with S_0
PRINT out the proportion of times S was larger than S_0
```

```
REARRANGE
```

```
Set s to the number of observations in the block
Start: Choose a random integer k from 0 to s-1
  Swap X[k] and X[s-1]:
    Decrement s and repeat from start
    Stop after you've selected all but one of the samples.
```

```
GET_DATA
user-written procedure gets data and packs it into a
two-dimensional array in which each row corresponds to
a block.
```

```
COMPUTE
F_1 = sum_{i=1}^2 sum_{j=1}^3|X_{ij.}-X_{.j.}|
for each block
  calculate the mean of that block
  for each level within a block
    calculate the mean of that block-level
    calculate difference from block mean
```

Estimating Significance Level of the Interaction of Sunlight and Fertilizer on Crop Yield

Test statistic based on the deviates from the additive model. Design must be balanced.

Set aside space for

Monte	the number of Monte Carlo simulations
S_0	the original value of test statistic
S	test statistic for rearranged data
data	{5, 10, 8, 15, 22, 18, 21, 29, 25, 6, 9, 12, 25, 32, 40, 55, 60, 48};
deviates	vector of deviates
$n = 3$	number of observations in each category
blocks $= 2$	number of blocks
levels $= 3$	number of levels of factor

```
MAIN program
GET_DATA
Calculate the DEVIATES
COMPUTE the test statistic S_0
Repeat Monte times:
   REARRANGE the observations
  COMPUTE the test statistic S
  Compare S with S_0
Print out the proportion of times S was larger than S_0
```

```
COMPUTE sum_i^I sum_j^J (sum_k^K X'_{ik})^2
  for each block
```

```
for each level
  sum the deviates
  square this sum
  cumulate
```

DEVIATES X'_{ijk} = X_{ijk}-X_{i..} - X_{.j.} + X_{...}
Set aside space for level means, block means, and grand mean
for each level calculate mean
for each block
  calculate mean
  for each level
    cumulate grand mean
for each block
  for each level
    calculate deviate from additive model

Bounds on Significance Level of Type I Censored Data

Set aside space for

Monte		the number of Monte Carlo simulations
$T_o = S_U(o) + CN_C(o)$		the original value of the test statistic
$T = S_U(\pi) + CN_C(\pi)$		test statistic for rearranged data
data		$\{880, 3430, 2860, C, 4750, 3820, C, 7170, C, 4800\}$;
$n = 5, m = 5$		number of observations in each category
$C = 10,000$		

Main program:
GET_DATA
  put all the observations into a single linear vector
COMPUTE T_o for the original observations
Repeat Monte times:
  REARRANGE the data in the block
  COMPUTE S_U, N_C, T
  Compare T with T_0
PRINT out the proportion of times T was larger than T_0

REARRANGE
Start: Set s=n
  Choose a random integer k from 0 to s-1
  Swap X[k] and X[s-1]:
    Decrement s and repeat from start
  Stop when s=0.

GET_DATA
user-written procedure gets data and packs it into a
linear vector

```
COMPUTE
N_C = S_U =0
For i=1 to n
  If X[i] = C then N_C++
    else S_U + = X[i]
T = S_U + CN_C
```

Two-Sample Test for Dispersion Using Aly's Statistic Written in SC^{TM} (see Appendix 4)

```
good_aly(X,Y,s)
  X,Y -> data vectors, size nx,ny
  s -> no. of assignments to sample (default: all)
```

If $nx = ny$, $m = nx$, otherwise let m be the smaller of nx, ny and replace the larger vector by a sample of size m from it. Let x, y be the corresponding order statistics. Let $\{D_x, D_y\}$ be the $m - 1$ differences $x[i + 1] - x[i]$ and $y[i + 1] - y[i]$, $i = 1, \ldots, m - 1$, and let $w[i] = i(m - i)$. These $\{D_x, D_y\}$ are passed to perm2_(), which computes $w.\pi(D_x)$ where $\pi(D_x)$ is a random rearrangement of m elements from $\{D_x, D_y\}$ and finds the two-tailed fraction of rearrangements for which $|w.D_x| \geq |w.\pi(D_x)|$.

```
func good_aly_f(){
  sort($1$)
  return g_c*$1$
}


proc good_aly(){
  local(ac,el,obs,X,Y,m,mm1,m1)

  # check arguments
  if(ARGCHECK){
    m1= ''(vec,vec) or (vec,vec,expr)\n''
    if(NARGS<2 || NARGS>3) abort_(ui,$0$,m1)
    if(!isvec($1$) || !isvec($2$)) abort_(ui,$0$,m1)
    if(NARGS>=3){
      if($3$!=nint($3$) || $3$<10) abort_($0$,a3od)
    }
  }

  # set 1-up element-numbering
  el= ELOFFSET; ELOFFSET= -1
  # arrive at, if necessary, equal-size (m) vectors
  X= $1$; Y= $2$
  if(sizeof(X)<sizeof(Y)){
    shuffle(Y); limit(Y,1,m=sizeof(X))
  }else if(sizeof(Y)<sizeof(X)){
```

```
   shuffle(X); limit(X,1,m=sizeof(Y))
}else m= sizeof(X)
mm1= m - 1

# get the pairwise differences
sort(X); X= X[2:m]-X[1:mm1]
sort(Y); Y= Y[2:m]-Y[1:mm1]
# get the multipliers for the differences, as global g_c
vector(g_c,mm1)
ords(g_c)
g_c= g_c *_(-g_c+m)
# Result for the actual data:-
print(''\tAly delta='',obs=(good_aly_f(Y) -
good_aly_f(X))/(m*m))

# Call perm2_() to give the left and right tail over all
  # (3 arguments) or some (4 arguments) using the statistic
  # returned by good_aly_f(): order or arguments is reversed
  # in perm2_() calls to make scale(Y) > scale(X) correspond
  # to a small upper tail probability, for consistency with
  # other similar sc routines

  # turn off argument-checking for the call to perm2_()
  ac= ARGCHECK; ARGCHECK= 0

  if(NARGS==2){
    # complete enumeration
    perm2_(good_aly_f,Y,X)
  }else{
    # subset only
    perm2_(good_aly_f,Y,X,$3$)
  }

  # free globals
  free(g_c)

  # restore re-set globals
  AUX[5]= obs
  ARGCHECK= ac
  ELOFFSET= el
}
```

Using Stata for Permutation Tests[1]

Put the following program in a file called **permute.ado**. Be sure that the file is saved as a plain text (ASCII) file and that there is a hard return at the end of the last line in the file.

Put **permute.ado** in your **C:\ADO** directory (you may have to create this directory) or your Stata working directory (i.e., your current directory). Do not put it in the **C:\STATA\ADO** directory.

```
program define permute
  version 4.0
  parse '''*''', parse('' ,'')
  local prog '''1'''
  capture which 'prog'
  if _rc {
    di in red ''program 'prog' not found''
  exit 199
  }
  mac shift

  local varlist ''req ex''
  #delimit ;
  local options ''BY(str) LEft RIght Reps(int 100)
    DIsplay(int 10) EPS(real 1e-7) noProb POST(str) DOuble
    EVery(str) REPLACE'' ;
  #delimit cr
  parse '''*'''
  parse '''varlist''', parse('' '')
  local x '''1'''
  macro shift

  if '''by'''!='''' {
    unabbrev 'by'
    local by ''$S_1''
    local byby ''by 'by':''
  }
  if 'eps' < 0 {
    di in red ''eps() must be greater than or equal to zero''
    exit 198
  }
  if '''left'''!='''' & '''right'''!='''' {
    di in red ''only one of left or right can be specified''
```

---

[1]These ado files are authored by William M. Sribney, Stata Corporation.
(c) 1996 Stata Corporation, 702 University Drive East,
College Station, TX 77840, (409) 696-4600, (800) 782-8272,
Email: stata@stata.com, Web: http://www.stata.com.

```
      exit 198
   }
   if '''left'''!='''' {
     *local ho ''Test of Ho: T = 0 vs. Ha: T < 0 (one-sided)''
     local ho ''p is an estimate of Pr(T <= T(obs))''
     local rel ''<=''
     local eps = -'eps'
   }
   else if '''right'''!='''' {
     *local ho ''Test of Ho: T = 0 vs. Ha: T > 0 (one-sided)''
     local ho ''p is an estimate of Pr(T >= T(obs))''
     local rel ''>=''
   }
   else {
     *local ho ''Test of Ho: T = 0 vs. Ha: T sim= 0
         (two-sided)''
     local ho ''p is an estimate of Pr(|T| >= |T(obs)|)''
     local rel ''>=''
     local abs ''abs''
   }
   if '''post'''=='''' & '''double'''!='''' {
     di in red ''double can only be specified when using
         post()''
     exit 198
   }
   if '''post'''=='''' & '''every'''!='''' {
     di in red ''every() can only be specified when using
         post()''
     exit 198
   }
   if '''post'''=='''' & '''replace'''!='''' {
     di in red ''replace can only be specified when using
         post()''
     exit 198
   }

/* Get value of test statistic for unpermuted data. */

   preserve
   global S_1 ''first''

   if '''prob'''=='''' | '''post'''!='''' {
     capture noisily 'prog' 'x' '*'
     if _rc {
       di in red _n '''prog' returned error:''
       error _rc
```

```
    }
    if ``$S_1''==``first'' | ``$S_1''==`''' {
      di in red ```prog' does not set global macro S_1''
      exit 7
    }
    capture confirm number $S_1
    if _rc {
      di in red ```prog' returned '$S_1' where number '' /*
      */ ``expected''
      exit 7
    }
  }

/* Initialize postfile. */

  if ```post'''!=`''' {
    tempname postnam
    if ```every'''!=`''' {
      confirm integer number `every'
      local every ``every(`every')''
    }
    postfile `postnam' stat using `post', /*
    */ `replace' `double' `every'
  }
  else local po ``*''

/* Display observed test statistic and Ho. */

  set more 1
  di in gr ``(obs='' _N ``)''

  if ```prob'''==`''' {
    tempname comp
    local tobs ``$S_1''
    scalar `comp' = `abs'($S_1) - `eps'
    local dicont ``_c''
    local c 0
    di _n in gr ``Observed test statistic = T(obs) = '' /*
    */ in ye %9.0g `tobs' _n(2) in gr ```ho''' _n
  }
  else local so ``*''

/* Sort by `by' if necessary. */

  if ```by'''!=`''' { sort `by' }
```

```
/* Check if 'x' is a single dichotomous variable. */

  quietly {
    tempvar k
    summarize 'x'
    capture assert _result(1)==_N /*
    */ & ('x'==_result(5) | 'x'==_result(6))
    if _rc==0 {
      tempname min max
      scalar 'min' = _result(5)
      scalar 'max' = _result(6)

      'byby' gen long 'k' = sum('x'=='max')
      'byby' replace 'k' = 'k'[_N]

      local oo ''*''
    }
    else {
      gen long 'k' = _n
      local do ''*''
    }
  }

/* Do permutations. */

  local i 1
  while 'i' <= 'reps' {
    'oo' PermVars '''by''' 'k' 'x'

    'do' PermDiV '''by''' 'k' 'min' 'max' 'x'

    'prog' 'x' '*'
    'so' local c = 'c' + ('abs'($S_1) 'rel' 'comp')
    'po' post 'postnam' $S_1

    if 'display' > 0 & mod('i','display')==0 {
      di in gr ''n = '' in ye %5.0f 'i' 'dicont'
      'so' noi di in gr '' p = '' /*
      */ in ye %4.0f 'c' ''/'' %5.0f 'i' /*
      */ in gr '' = '' in ye %7.5f 'c'/'i' /*
      */ in gr '' s.e.(p) = '' in ye %7.5f /*
      */ sqrt(('i'-'c')*'c'/'i')/'i'
    }
    local i = 'i' + 1
  }
```

```
  if '''prob'''=='''' {
    if 'display' > 0 { di }
    di in gr ''n = '' in ye %5.0f 'reps' /*
    */ in gr '' p = '' in ye %4.0f 'c' ''/'' %5.0f 'reps' /*
    */ in gr '' = '' in ye %7.5f 'c'/'reps' /*
    */ in gr '' s.e.(p) = '' in ye %7.5f /*
    */ sqrt(('reps'-'c')*'c'/'reps')/'reps'
  }

  if '''post'''!='''' { postclose 'postnam' }

/* Save global macros. */

  global S_1 '''reps'''
  global S_2 '''c'''
  global S_3 '''tobs'''
end

program define PermVars /* ''byvars'' k var */
  version 4.0
  local by '''1'''
  local k '''2'''
  local x '''3'''
  tempvar r y
  quietly {
    if '''by'''!='''' {
      by 'by': gen double 'r' = uniform()
    }
    else gen double 'r' = uniform()

    sort 'by' 'r'
    local type : type 'x'
    gen 'type' 'y' = 'x'['k']
    drop 'x'
    rename 'y' 'x'
  }
end

program define PermDiV /* ''byvars'' k min max var */
  version 4.0
  local by '''1'''
  local k '''2'''
  local min '''3'''
  local max '''4'''
  local x '''5'''
  tempvar y
```

```
if '''by'''!=''' {
   sort 'by'
   local byby ''by 'by':''
}
quietly {
   gen byte 'y' = . in 1
   'byby' replace 'y' =
uniform()<(('k'-sum('y'[_n-1]))/(_N-_n+1)
   replace 'x' = cond('y','max','min')
}
end
```

progname is the name of a program that computes the test statistic and places its value in the global macro ^S_1^. The arguments to **progname** are *varname1* [*varlist*]. For each repetition, the values of *varname1* are randomly permuted, **progname** is called to compute the test statistic, and a count is kept whether this value of the test statistic is more extreme than the observed test statistic. The values of the test statistic for each random permutation can also be stored in a data set using the ^post()^ option.

Options

^by(^groupvars^)^ specifies that the permutations be performed within each group defined by the values of groupvars; i.e., group membership is fixed and the values of *varname1* are independently permuted within each group. For example, this permutation scheme is used for randomized-block ANOVA to permute values within each block.

^reps(^#^)^ specifies the number of random permutations to perform. Default is 100.

^display(^#^)^ displays output every #-th random permutation. Default is 10. ^display(0)^ suppresses all but the final output.

^left^ | ^right^ request that one-sided p-values be computed. If ^left^ is specified, an estimate of $Pr(T <= T(\text{obs}))$ is produced, where T is the test statistic and $T(\text{obs})$ is its observed value. If ^right^ is specified, an estimate of $Pr(T >= T(\text{obs}))$ is produced. By default, two-sided p-values are computed; i.e., $Pr(|T| >= |T(\text{obs})|)$ is estimated.

^noprob^ specifies that no p-values are to be computed.

^eps(^#^)^ specified the numerical tolerance for testing $|T| >= |T(\text{obs})|$, $T <= T(\text{obs})$, or $T >= T(\text{obs})$. These are considered true if, respectively, $|T| >= |T(\text{obs})| - \#$, $T <= T(\text{obs}) + \#$, or $T >= T(\text{obs}) - \#$. By default, it is $1e - 7$. ^eps()^ should not have to be set under normal circumstances.

^post(^filename^)^ specifies a name of a ^.dta^ file that will be created holding the values of the test statistic computed for each random permutation.

^double^ can only be specified when using ^post()^. It specifies that the values of the test statistic be stored as type ^double^; default is type ^float^.

ˆevery(ˆ#ˆ)ˆ can only be specified when using ˆpost()ˆ. It specifies that the values of test statistic be saved to disk every #-th repetition.

ˆreplaceˆ indicates that the file specified by ˆpost()ˆ may already exist and, if it does, it can be erased and replaced by a new one.

Note: ˆpermuteˆ works faster when *varname1* is a 0/1 variable (with no missing values). So, if using a 0/1 variable, specify it as the one to be permuted.

progname must have the following outline:

```
program define progname
  compute test statistic
  global S_1 = test statistic
end
```

Arguments to **progname** are *varname1* [*varlist*]; i.e., the same variables that specified with ˆpermuteˆ are passed to **progname**. Here is an example of a program that estimates the permutation distribution *p*-value for the Pearson correlation coefficient:

```
program define permpear
  quietly corr '1' '2'
  global S_1 = _result(4)
end
```

To use this program, call ˆpermuteˆ using

. ˆpermute permpear x yˆ

In addition, the global macro S_1 is set to "first" for the first call to **progname**, which computes the observed test statistic $T(obs)$; i.e., $T(obs)$ is the value of the test statistic for the unpermuted data.

Thus, **progname** can optionally have the form:

```
program define progname /* args = varname1 [varlist] */
if ''$S_1'' == ''first'' {
do initial computations
}
compute test statistic
global S_1 = test statistic
end
```

Here is an example of a program that estimates the permutation p-value for the two-sample *t* test:

```
program define permt2
local grp '''1'''
local x '''2'''
tempvar sum
quietly {
```

```
  if ''$S_1''==''first'' {
    gen double 'sum' = sum('x')
    scalar _TOTAL = 'sum'[_N]
    drop 'sum'
    summarize 'grp'
    scalar _GROUP1 = _result(5)
    count if 'grp'==_GROUP1
    scalar _TOTAL = (_result(1)/_N)*_TOTAL
  }
  gen double 'sum' = sum(('grp'==_GROUP1)*'x')
  global S_1 = 'sum'[_N] - _TOTAL
}
end
```

To use this program, call ^permute^ using

```
. ^permute permt2 group x^
```

Examples include

```
. ^permute permpear x y^
. ^permute permpear x y, reps(1000)^
. ^permute permpear x y, reps(10000) display(100)^
. ^permute permpear x y, reps(1000) di(100) post(pearson)^
. ^permute permpear x y, reps(10000) di(1000) post(pearson)
        replace every(1000) double^
. ^permute permt2 group x^
. ^permute permt2 group x, left^
. ^permute panova treat outcome subject, by(subject)
        reps(1000)^
```

```
program define permpear
  version 4.0
  qui corr '1' '2'
  global S_1 = _result(4)
end
```

```
program define permt2
  version 4.0
  local grp '''1'''
  local x '''2'''
  tempvar sum
  quietly {
    if ''$S_1''==''first'' {
    gen double 'sum' = sum('x')
    scalar _TOTAL = 'sum'[_N]
    drop 'sum'
    summarize 'grp'
```

```
      scalar _GROUP1 = _result(5)
      count if 'grp'==_GROUP1
      scalar _TOTAL = (_result(1)/_N)*_TOTAL
   }
   gen double 'sum' = sum(('grp'==_GROUP1)*'x')
   global S_1 = 'sum'[_N] - _TOTAL
}
end

program define panova
   version 4.0
   local treat '''1'''
   local x '''2'''
   local subj '''3'''
   quietly {
      anova 'x' 'subj' 'treat'
      test 'treat'
      global S_1 = _result(6)
   }
end
```

To Learn More

Consult Good [1994].

SAS and S-PLUS Code for Bootstraps

S-PLUS Code

SPLUS code for obtaining BC_a intervals may be obtained from the statistics archive of Carnegie-Mellon University by sending an email to **statlib@lib.stat.cmu** with the one-line message **snd bootstrap.funs from** S. Hints on usage will be found in Efron and Tibshirani [1993].

Stata Code

Bootstrap Samples of Correlation

```
program define bcorr
  tempname sim
    postfile 'sim' rho using results, replace
    quietly {
      local i =1
      while 'i' <= 200 {
        use mardia.dta, clear
        bsample _N
        correlate alg stat
        post 'sim' _result(4)
        local i = 'i' + 1
        }
      }
    postclose 'sim'
end
bcorr
```

```
use results.dta, clear
summarize, detail
```

SAS General-Purpose Macros

SAS macros for density estimation may be found in *SAS System for Graphics* [1991] (pp. 116–7, 556–558) and are the work of Michael Friendly, York University, Ontario, Canada.

```
%macro boot(          /* Bootstrap resampling analysis */
data=,                /* Input data set, not a view or a tape
                         file. */
samples=200,          /* Number of resamples to generate. */
residual=,            /* Name of variable in the input data
                         set that contains residuals; may not
                         be used with SIZE= */
equation=,            /* Equation (in the form of an
                         assignment statement) for computing
                         the response variable */
size=,                /* Size of each resample; default is
                         size of the input data set. The SIZE=
                         argument may not be used with
                         BALANCED=1 or with a nonblank value
                         for RESIDUAL= */
balanced=,            /* 1 for balanced resampling; 0 for
                         uniform resampling. By default,
                         balanced resampling is used unless
                         the SIZE= argument is specified, in
                         which case uniform resampling is
                         used. */
random=0,             /* Seed for pseudorandom numbers. */
stat=_numeric_,       /* Numeric variables in the OUT= data
                         set created by the %ANALYZE macro
                         that contain the values of statistics
                         for which you want to compute
                         bootstrap distributions. */
id=,                  /* One or more numeric or character
                         variables that uniquely identify the
                         observations of the OUT= data set
                         within each BY group. No ID variables
                         are needed if the OUT= data set has
                         only one observation per BY group.
                         The ID variables may not be named
                         _TYPE_, _NAME_, or _STAT_*/
biascorr=1,           /* 1 for bias correction; 0 otherwise */
```

```
alpha=.05,              /* significance (i.e., one minus
                           confidence) level for confidence
                           intervals; blank to suppress normal
                           confidence intervals */
print=1,                /* 1 to print the bootstrap estimates;
                           0 otherwise. */
chart=1                 /* 1 to chart the bootstrap resampling
                           distributions; 0 otherwise. */
  );
+by=0;
  %global usevardf vardef;
  %let usevardf=0;

  *** compute the actual values of the statistics;
  %let vardef=DF;
  %let by=;
  %analyze(data=&data,out=_ACTUAL_);
  %if &syserr>4 %then %goto exit;

  *** compute plug-in estimates;
  %if &usevardf %then %do;
    %let vardef=N;
    %analyze(data=&data,out=_PLUGIN_);
    %let vardef=DF;
    %if &syserr>4 %then %goto exit;
  %end;

  %if &useby=0 %then %let balanced=0;

  %if %bquote(&size)^= %then %do;
    %if %bquote(&balanced)= %then %let balanced=0;
    %else %if &balanced %then %do;
      %put %cmpres(ERROR in BOOT: The SIZE= argument may not
        be used with BALANCED=1.);
      %goto exit;
    %end;
    %if %bquote(&residual)^= %then %do;
      %put %cmpres(ERROR in BOOT: The SIZE= argument may not
        be used with RESIDUAL=.);
      %goto exit;
    %end;
  %end;
  %else %if %bquote(&balanced)= %then %let balanced=1;

  *** find number of observations in the input data set;
  %global _nobs;
```

```
data _null_;
  call symput('_nobs',trim(left(put(_nobs,12.))));
  if 0 then set &data nobs=_nobs;
  stop;
run;
%if &syserr>4 %then %goto exit;

%if &balanced %then
  %bootbal(data=&data,samples=&samples,
    random=&random,print=0);
  %else %if &useby %then
    %bootby(data=&data,samples=&samples,
      random=&random,size=&size,print=0);

%if &syserr>4 %then %goto exit;

%if &balanced | &useby %then %do;
  %let by=_sample_;
  %analyze(data=BOOTDATA,out=BOOTDIST);
%end;

%else
  %bootslow(data=&data,samples=&samples,
    random=&random,size=&size);

%if &syserr>4 %then %goto exit;

%if &chart %then %do;
  %if %bquote(&id)^= %then %do;
    proc sort data=BOOTDIST; by &id; run;
    proc chart data=BOOTDIST(drop=_sample_);
      vbar &stat;
      by &id;
    run;
  %end;
  %else %do;
    proc chart data=BOOTDIST(drop=_sample_);
      vbar &stat;
    run;
  %end;
%end;

%bootse(stat=&stat,id=&id,alpha=&alpha,biascorr=&biascorr,
  print=&print)

%exit:;
```

```
%mend boot;

%macro bootbal( /* Balanced bootstrap resampling */
  data=&_bootdat,
  samples=200,
  random=0,
  print=0,
  );
  * Gleason, J.R. (1988) ''
  data BOOTDATA/view=BOOTDATA;
    %bootin;
    drop _a _cbig _ii _j _jbig _k _s;
    array _c(&_nobs) _temporary_; /* cell counts */
    array _p(&_nobs) _temporary_; /* pointers */
    do _j=1 to &_nobs;
      _c(_j)=&samples;
    end;
    do _j=1 to &_nobs;
      _p(_j)=_j;
    end;
    _k=&_nobs; /* number of nonempt y cells left */
    _jbig=_k; /* index of largest cell */
    _cbig=&samples; /* _cbig >= _c(_j)
*/
    do _sample_=1 to &samples;
      do _i=1 to &_nobs;
        do until(_s<=_c(_j));
          _j=ceil(ranuni(&random)*_k); /* choose a cell */
          _s=ceil(ranuni(&random)*_cbig); /* accept cell? */
        end;
        _l=_p(_j);
        _obs_=_l;
        _c(_j)+-1;
* put _sample_= _i= _k= _l= @30 %do i=1 %to &_nobs; _c(&i)
  %end;;
        if _j=_jbig then do;
          _a=floor((&samples-_sample_-_k)/_k);
          if _cbig-_c(_j)>_a then do;
            do _ii=1 to _k;
              if _c(_ii)>_c(_jbig) then
_jbig=_ii;
            end;
            _cbig=_c(_jbig);
          end;
        end;
```

```
         if _c(_j)=0 then do;
           if _jbig=_k then _jbig=_j;
           _p(_j)=_p(_k);
           _c(_j)=_c(_k);
           _k+-1;
         end;
         %bootout(_l);
       end;
     end;
     stop;
   run;
   %if &syserr>4 %then %goto exit;

   %if &print %then %do;
     proc print data=BOOTDATA; id _sample__obs_; run;
   %end;

%exit:;

%mend bootbal;

%macro bootby( /* Uniform bootstrap resampling */
   data=&_bootdat,
   samples=200,
   random=0,
   size=,
   print=0
   );

   %if %bquote(&size)= %then %let size=&_nobs;

   data BOOTDATA/view=BOOTDATA;
     %bootin;
     do _sample_=1 to &samples;
       do _i=1 to &size;
         _p=ceil(ranuni(&random)*&_nobs);
         _obs_=_p;
         %bootout(_p);
       end;
     end;
     stop;
   run;
   %if &syserr>4 %then %goto exit;

   %if &print %then %do;
     proc print data=BOOTDATA; id _sample__obs_; run;
```

```
  %end;

%exit:;
%mend bootby;

%macro bootslow( /* Uniform bootstrap resampling and
         analysis without BY processing */
  data=&_bootdat,
  samples=20,
  random=0,
  size=
  );

  %put %cmpres(WARNING: Bootstrap analysis will be slow
    because the sisi ANALYZE macro did not use the BYSTMT
    macro.);

  %if %bquote(&size)= %then %let size=&_nobs;

  data BOOTDIST; set _ACTUAL_; _sample_=0; delete; run;
  options nonotes;
  %local sample;
  %do sample=1 %to &samples;
    %put Bootstrap sample &sample;
    data _TMPD_;
      %bootin;
      do _i=1 to &size;
        _p=ceil(ranuni(%eval(&random+&sample))*&_nobs);
        %bootout(_p);
      end;
      stop;
    run;
    %if &syserr>4 %then %goto exit;

    %analyze(data=_TMPD_,out=_TMPS_);
    %if &syserr>4 %then %goto exit;
    data _TMPS_; set _TMPS_; _sample_=&sample; run;
    %if &syserr>4 %then %goto exit;
    proc append data=_TMPS_base=BOOTDIST; run;
    %if &syserr>4 %then %goto exit;
  %end;

%exit:;
  options notes;
%mend bootslow;
```

Confidence Intervals for the Correlation Coefficient

SAS Code to Obtain the Bootstrap Distribution

The following %ANALYZE macro can be used to process all the statistics in the OUT= data set from PROC CORR: it extracts only the relevant observations and variables and provides descriptive names and labels.

```
%macro analyze(data=,out=);
   proc corr noprint data=&data out=&out;
      var lsat gpa;
      %bystmt;
   run;
   %if &syserr=0 %then %do;
      data &out;
         set &out;
         where _type_='CORR' & _name_='LSAT';
         corr=gpa;
         label corr='Correlation';
         keep corr &by;
      run;
   %end;
%mend;

title2 'Bootstrap Analysis';
%boot(data=law,random=123)
```

```
%macro bootci(          /* Bootstrap percentile-based confidence
                           intervals. Creates output data set
                           BOOTCI. */
method,                 /* One of the following methods must be
                           specified:
                              PERCENTILE or PCTL
                              HYBRID
                              T
                              BC
                              BCA       Requires the %JACK macro */
stat=,                  /* Numeric variables in the OUT= data set
                           created by the %ANALYZE macro that
                           contain the values of statistics for
                           which you want to compute bootstrap
                           distributions. */
student=,               /* For the T method only, numeric
                           variables in the OUT= data set created
                           by the %ANALYZE macro that contain the
                           standard errors of the statistics for
```

which you want to compute bootstrap
distributions. There must be a
one-to-one between the VAR= variables
and the STUDENT= variables */
id=,   /* One or more numeric or character
variables that uniquely identify the
observations of the OUT= data set
within each BY group. No ID variables
are needed if the OUT= data set has
only one observation per BY group. The
ID variables may not be named _TYPE_,
NAME, or _STAT_*/
alpha=.05,   /* significance (i.e., one minus
confidence) level for confidence
intervals */
print=1);   /* 1 to print the bootstrap confidence
intervals; 0 otherwise. */

```
%global _bootdat;
%if %bquote(&_bootdat)= %then %do;
  %put ERROR in BOOTCI: You must run BOOT before BOOTCI;
  %goto exit;
%end;

*** check method;
data _null_;
  length method $10;
  method=upcase(symget('method'));
  if method=' ' then do;
    put 'ERROR in BOOTCI: You must specify one of the
        methods ' 'PCTL, HYBRID, T, BC or BCa';
    abort;
  end;
  else if method='PERCENTILE' then method='PCTL';
  else if method not in ('PCTL' 'HYBRID' 'BC' 'BCA' 'T')
    then do;
    put ''ERROR in BOOTCI: Unrecognized method '''
        method '''';
    abort;
  end;
  call symput('qmethod',method);
run;
%if &syserr>4 %then %goto exit;

%if &qmethod=T %then %do;
  %if %bquote(&stat)= | %bquote(&student)= %then %do;
```

```
      data _null_;
put 'ERROR: VAR= and STUDENT= must be specified with the
           T method';
      run;
      %goto exit;
    %end;
  %end;


  *** sort resampling distributions;
  %if %bquote(&id)^= %then %do;
    proc sort data=BOOTDIST;
      by &id _sample_;
    run;
    %if &syserr>4 %then %goto exit;
  %end;


  *** transpose resampling distributions;
  proc transpose data=BOOTDIST prefix=col
    out=BOOTTRAN(rename=(col1=value _name_=name));
    %if %bquote(&stat)^= %then %do;
      var &stat;
    %end;
    by %if %bquote(&id)^= %then &id; _sample_;
  run;
  %if &syserr>4 %then %goto exit;


  %if &qmethod=T %then %do;
    *** transpose studentizing statistics;
    proc transpose data=BOOTDIST prefix=col
      out=BOOTSTUD(rename=(col1=student _name_=studname));
        var &student;
      by %if %bquote(&id)^= %then &id; _sample_;
    run;
    %if &syserr>4 %then %goto exit;


    data BOOTTRAN;
      merge BOOTTRAN BOOTSTUD;
      label student='Value of Studentizing Statistic'
        studname='Name of Studentizing Statistic';
    run;
    %if &syserr>4 %then %goto exit;
  %end;


  proc sort data=BOOTTRAN;
    by
      %if %bquote(&id)^= %then &id;
```

```
      name
      %if &qmethod=BC | &qmethod=BCA %then value;
      %else %if &qmethod=T %then _sample_;
    ;
  run;
  %if &syserr>4 %then %goto exit;

  %if &qmethod=T %then %do;
    *** transpose the actual statistics in each observation
      must get data set in unsorted order for merge;
    proc transpose data=_ACTUAL_out=_ACTTR_prefix=value;
      %if %bquote(&stat)^= %then %do;
      var &stat;
    %end;
    %if %bquote(&id)^= %then %do;
      by &id;
    %end;
  run;
  %if &syserr>4 %then %goto exit;

  *** transpose the actual studentizing statistics;
  proc transpose data=_ACTUAL_prefix=col
      out=_ACTSTUD(rename=(_name_=studname col1=student));
      var &student;
    %if %bquote(&id)^= %then %do;
      by &id;
    %end;
  run;
  %if &syserr>4 %then %goto exit;

  *** merge statistics with studentizing statistics;
  data _ACT_T_;
    merge _ACTTR__ACTSTUD;
    label student='Value of Studentizing Statistic'
      studname='Name of Studentizing Statistic';
  run;
  %if &syserr>4 %then %goto exit;

  proc sort data=_ACT_T_;
    by %if %bquote(&id)^= %then &id; _name_;
  run;
  %if &syserr>4 %then %goto exit;

  data BOOTTRAN;
    merge BOOTTRAN _ACT_T_(rename=(_name_=name));
    by
```

```
    %if %bquote(&id)^= %then &id;
    name
  ;
  value=(value-value1)/student;
run;
%if &syserr>4 %then %goto exit;
%end;


%if &qmethod=BC | &qmethod=BCA %then %do;

  %if &qmethod=BCA %then %do;
    %global _jackdat;
    %if %bquote(&_jackdat)^=%bquote(&_bootdat)
%then %do;
      %jack(data=&_bootdat,stat=&stat,id=&id,alpha=&alpha,
        chart=0,print=&print);
      %if &syserr>4 %then %goto exit;
    %end;

    *** estimate acceleration for BCa;
    proc means data=JACKDIST noprint vardef=df;
      %if %bquote(&stat)^= %then %do;
        var &stat;
      %end;
      output out=JACKSKEW(drop=_type__freq_
_sample_) skewness=;
      %if %bquote(&id)^= %then %do;
        by &id;
      %end;
    run;
    %if &syserr>4 %then %goto exit;

    *** transpose skewness;
    proc transpose data=JACKSKEW prefix=col
      out=_ACCEL_(rename=(col1=skewness _name_=name));
      %if %bquote(&stat)^= %then %do;
        var &stat;
      %end;
      %if %bquote(&id)^= %then %do;
        by &id;
      %end;
    run;
    %if &syserr>4 %then %goto exit;

    proc sort data=_ACCEL_;
      by %if %bquote(&id)^= %then &id; name ;
```

```
      run;
      %if &syserr>4 %then %goto exit;
    %end;

    *** estimate median bias for BC;
    data _BC_;
      retain _alpha _conf;
      drop value value1;
      if _n_=1 then do;
        _alpha=&alpha;
        _conf=100*(1-_alpha);
        call symput('conf',trim(left(put(_conf,best8.))));
      end;

      merge _ACTTR_(rename=(_name_=name))
          BOOTTRAN;
      by %if %bquote(&id)^= %then &id; name;
      if first.name then do; n=0; _z0=0; end;
      n+1;
      _z0+(value<value1)+.5*(value=value1);
      if last.name then do;
        _z0=probit(_z0/n);
        output;
      end;
    run;
    %if &syserr>4 %then %goto exit;

    *** compute percentiles;
    data BOOTPCTL;
      retain _i _lo _up _nplo _jlo _glo _npup _jup _gup
          alcl aucl;
      drop _alpha _sample__conf _i _nplo _jlo _glo
_npup _jup _gup
          value;
      merge BOOTTRAN _BC_%if &qmethod=BCA %then _ACCEL_;;
      by %if %bquote(&id)^= %then &id; name;
      label _lo='Lower Percentile Point'
        _up='Upper Percentile Point'
        _z0='Bias Correction (Z0)';
      if first.name then do;
        %if &qmethod=BC %then %do;
          _lo=probnorm(_z0+(_z0+probit(_alpha/2)));
          _up=probnorm(_z0+(_z0+probit(1-_alpha/2)));
        %end;
        %else %if &qmethod=BCA %then %do;
          drop skewness;
```

```
      retain _accel;
      label _accel='Acceleration';
      _accel=skewness/(-6*sqrt(&_nobs))*
        (&_nobs-2)/&_nobs/sqrt((&_nobs-1)/&_nobs);
      _i=_z0+probit(_alpha/2);
      _lo=probnorm(_z0+_i/(1-_i*_accel));
      _i=_z0+probit(1-_alpha/2);
      _up=probnorm(_z0+_i/(1-_i*_accel));
    %end;
    _nplo=min(n-.5,max(.5,fuzz(n*_lo)));
    _jlo=floor(_nplo); _glo=_nplo-_jlo;
    _npup=min(n-.5,max(.5,fuzz(n*_up)));
    _jup=floor(_npup); _gup=_npup-_jup;
    _i=0;
  end;
  _i+1;
  if _glo then do;
    if _i=_jlo+1 then alcl=value;
  end;
  else do;
    if _i=_jlo then alcl=value;
    else if _i=_jlo+1 then alcl=(alcl+value)/2;
  end;
  if _gup then do;
    if _i=_jup+1 then aucl=value;
  end;
  else do;
    if _i=_jup then aucl=value;
    else if _i=_jup+1 then aucl=(aucl+value)/2;
  end;
  if last.name then do;
    output;
  end;
  run;
  %if &syserr>4 %then %goto exit;
%end;

%else %do;
  %local conf pctlpts pctlpre pctlname;
  %let pctlpre=a;
  %let pctlname=lcl ucl;
  data _null_;
    _alpha=&alpha;
    _conf=100*(1-_alpha);
    call symput('conf',trim(left(put(_conf,best8.))));
    %if &qmethod=PCTL %then %do;
```

```
      _lo=_alpha/2;
      _up=1-_lo;
    %end;
    %else %if &qmethod=HYBRID | &qmethod=T %then %do;
      _up=_alpha/2;
      _lo=1-_up;
    %end;
    _lo=100*_lo;
    _up=100*_up;
    call symput('pctlpts',trim(left(put(_lo,best8.)))||'
'||
         trim(left(put(_up,best8.)))));
  run;
  %if &syserr>4 %then %goto exit;

  proc univariate data=BOOTTRAN noprint pctldef=5;
      var value;
      output out=BOOTPCTL n=n
        pctlpts=&pctlpts pctlpre=&pctlpre pctlname=&pctlname;
      by %if %bquote(&id)^= %then &id; name;
    run;
    %if &syserr>4 %then %goto exit;
%end;

data BOOTCI;
  retain &id name value alcl aucl confid method n;
  merge
    %if &qmethod=T
      %then _ACT_T_(rename=(_name_=name value1=value));
      %else _ACTTR_(rename=(_name_=name value1=value));
    BOOTPCTL;
  by %if %bquote(&id)^= %then &id; name;
  %if &qmethod=HYBRID %then %do;
    aucl=2*value-aucl;
    alcl=2*value-alcl;
  %end;
  %else %if &qmethod=T %then %do;
    aucl=value-aucl*student;
  alcl=value-alcl*student;
  %end;
  confid=&conf;
  length method $20;
  method='Bootstrap '||symget('method');
  label name ='Name'
    value ='Observed Statistic'
    alcl ='Approximate Lower Confidence Limit'
```

```
      aucl ='Approximate Upper Confidence Limit'
      confid='Confidence Level (%)'
      method='Method for Confidence Interval'
      n ='Number of Resamples'
      ;
  run;
  %if &syserr>4 %then %goto exit;

  %if &print %then %do;
    proc print data=BOOTCI label;
      id %if %bquote(&id)^= %then &id; name;
    run;
  %end;

%exit:
%mend bootci;
```

APPENDIX 4

Resampling Software

DBMS Copy, though not a statistics package, is a must for anyone contemplating the use of resampling methods. DBMS Copy lets you exchange files among a dozen or more statistics packages, a half dozen database managers, and several spreadsheets. This book could not have been put together without its aid. DOS and Windows versions. Conceptual Software, 9660 Hillcroft #510, Houston TX 77096; 713/721-4200.

StatXact is for anyone who routinely analyzes categorical or ordered data. The StatXact manual is a textbook in its own right. Can perform any of the techniques outlined in Chapter 6. Versions for MS DOS, OS 2, Windows, or UNIX. Also available as an add-on module for both SAS and SPSS. **LogXact**, a second product, supports exact inference for logistic regression and provides conditional logistic regression for matched case-control studies or clustered binomial data. Cytel Software Corporation, 675 Massachusetts Avenue, Cambridge, MA 02139; 617/661-2011; e-mail: mehta@jimmy.harvard.edu

Blossom Statistical Analysis Package is an interactive program for analyzing data using multiresponse permutation procedures (MRPP); includes statistical procedures for grouped data, agreement of model predictions, circular-distributions, goodness of fit, least absolute deviation, and quantile regression. Programmed by Brian Cade at the U.S. Geological Survey, Midcontinent Ecological Science Center. PC only with online manual in HTML Frames format. Freeware; may be downloaded from http://www.mesc.usgs.gov/blossom/blossom.html.

CART provides all the flexibility of the CART methodology for discrimination, including a priori probabalities and relative costs. DOS and Windows. Salford Systems, 5952 Bernadette Lane, San Diego, CA 92120.

NPSTAT and **NPFACT** support 2- and 3-factor designs, repeated measures, and correlations. Contact Richard May. Department of Psychology, University of Victoria, Victoria BC, Canada V8W 3P5.

nQuery Advisor helps you determine sample size for 50+ design and analysis combinations. A challenge to use but the advice is excellent. Windows. Statistical Solutions, 60 State Street, Suite 700, Boston, MA 02109; 800/262-1171; email:info@statsolusa.com.

RT is based on the programs in Brian Manly's book *Randomization and Monte Carlo Methods in Biology*. For DOS only. Analysis of variance with one to three factors, with randomization of observations or residuals, simple and linear multiple regression, with randomization of observations or residuals, Mantel's test with up to 9 matrices of dependent variables, Mead's test and Monte Carlo tests for spatial patterns, tests for trend, autocorrelation, and periodicity in time series. Particularly valuable for block matrix classification. Western EcoSystem Technology Inc., 1402 S. Greeley Highway, Cheyenne, WY 82007; 307/634-1756.

SAS (Statistical Analysis System) now offers StatXact (less the manual) as an add-on. The statistical literature is filled with SAS macros for deriving bootstrap and density estimates, as in Appendix 3. Offers a wider choice of ready-to-go graphics options than any other package. DOS, Windows, MVS, CMS, VMS, UNIX. SAS Institute Inc., SAS Campus Drive, Cary, NC 27513.

SCRT (Single Case Randomization Tests). Alternating treatments designs, AB designs, and extensions (ABA, ABAB, etc.), and multiple baseline designs. Customized statistics may be built and selected as the criteria for the randomization test. Provides a nonparametric meta-analytic procedure. DOS only. Interuniversity Expertise Center ProGAMMA, P.O. Box 841, 9700 AV Groningen, The Netherlands.

Simstat helps you derive the empirical sampling distribution of 27 different estimators. (DOS). Shareware. Normand Pelandeau, 5000 Adam Street, Montreal QC, Canada H1V 1W5.

S-PLUS is a language, a programming environment, and a comprehensive package of subroutines for statistical and mathematical analysis. Once only for do-it-yourselfers, a dozen texts and hundreds of file repositories on the Internet now offer all the subroutines you'll need for bootstrap, density estimation, and graphics. For UNIX and Windows. Statistical Sciences, Division of MathSoft, Inc., 1700 Westlake Avenue North, Suite 500, Seattle, WA 98109; 800/569-0123; www:http://statsci.com, email:brian@statsci.com.

SPSS offers ease of use, extensive graphics, and add-on modules for exact tests and neural nets. A bootstrap subcommand provides bootstrap estimates of the standard errors and confidence limits for constrained nonlinear regression. (Windows). SPSS Inc., 444 North Michigan Avenue, Chicago, IL 60611; 312/329-2400.

Stata provides a complete set of graphic routines plus subroutines and pre-programmed macros for bootstrap, density estimation, and permutation tests. DOS, Windows, UNIX. Stata Corp, 702 University Drive East, College Station, TX 77840; 800/782-8272; email:stata@stata.com.

Statistical Calculator (SC) DOS, UNIX, T800 transputer. An extensible statistical environment, supplied with more than 1200 built-in (compiled C) and external (written in SC's C-like language) routines. Permutation-based methods for contingency tables (χ square, likelihood, Kendall S, Theil U, κ, τ, odds ratio), one and two-sample inference for both means and variances, correlation, and multivariate analysis (MV runs test, Boyett/Schuster, Hoetelling's T^2. Ready-made bootstrap routines for testing homoscedacity, detecting multimodality, plus general bootstrapping and jack-knifing facilities. Dr. Tony Dusoir, Mole Software, 23 Cable Road, Whitehead, Co. Antrim BT38 9PZ, N. Ireland; (+44) (0)960 378988; email: fbgj23@ujvax.ulster.ac.uk.

Testimate lacks the pull-down menus and prompts that Windows users have grown to rely on. Too few statistical routines for its price tag. (DOS only.) idv: Datenanalyse und Versuchsplanung, Wessobrunner Strabe 6, D-82131 Gauting, Munich, Germany; 89.850.8001; also, SciTech International, Inc., 2231 North Clybourn Avenue, Chicago, IL 60614-3011; 800/622.3345.

NoName also lacks the pull-down menus and prompts that Windows users have grown to rely on. But it can be used to analyze all the permutation tests and many of the simpler bootstraps described in this text. Best of all for the nonstatistician, it tries to guide you to the correct test. Shareware. May be downloaded from **http://users.oco.net/drphilgood**.

Bibliographical References

[1] Aalen OO. Nonparametric inference for a family of counting processes. *Ann. Statist.* 1978; 6: 701–26.

[2] Abel U; Berger J. Comparison of resubstitution, data splitting, the bootstrap and the jacknife as methods for estimating validity indices of new marker tests: A Monte Carlo study. *Biometric J.* 1986; 28: 899–908.

[3] Abramson IS. Arbitrariness of the pilot estimate in adaptive kernel methods. *J Multiv Anal.* 1982; 12: 562–7.

[4] Abramson IS. On bandwidth variation in kernel estimates—a square root law. *Ann. Statist.* 1982; 10: 1217–29.

[5] Adams DC; Anthony CD. Using randomization techniques to analyse behavioural data. *Animal Behaviour.* 1996; 51: 733–738.

[6] Agresti A. *Categorical Data Analysis.* New York: John Wiley & Sons; 1990.

[7] Agresti A. A survey of exact inference for contingency tables. *Stat. Sci.* 1992; 7: 131–177.

[8] Ahmad IA; Kochar SC. Testing for dispersive ordering. *Statist Probab Ltr.* 1989; 8: 179–185.

[9] Ahrens LH. Observations of the Fe-Si-Mg relationship in chondrites. *Geochimica et Cosmochimica Acta.* 1965; 29: 801–6.

[10] Aitchison J; Lauder IJ. Kernel density estimation compositional data. *Appl. Stat.* 1985; 34: 129–37.

[11] Akritas MG. Bootstrapping the Kaplan-Meier estimator. *JASA.* 1986; 81: 1032–8.

[12] Albers W; Bickel PJ & Van Zwet WR. Asymptotic expansions for the power of distribution-free tests in the one-sample problem. *Ann. Statist.* 1976; 4: 108–156.

[13] Alderson MR; Nayak RA. Study of space-time clustering in Hodgkin's disease in the Manchester Region. British J. Preventive and Social Medicine.1971; 25:168-73.

[14] Altman DG; Andersen PK. Bootstrap investigation of the stability of the Cox regression model. *Stat. Med.* 1989; 8: 771–83.

[15] Aly E-E AA. Simple tests for dispersive ordering. *Stat. Prob. Ltr.* 1990; 9: 323–5.

[16] Andersen PK; Borgan O; Gill R & Keiding N. Linear nonparametric tests for comparison of counting processes with applications to censored survival data. *Int. Stat. Rev.* 1982; 50: 219–58.

[17] Andersen PK; Borgan O; Gill RD & Keiding N. *Statistical Models Based on Counting Processes.* New York: Springer Verlag; 1993.

[18] Arndt S; Cizadlo T; Andreasen NC; Heckel D; Gold S; & Oleary DS. Tests for comparing images based on randomization and permutation methods. *J. Cerebral Blood Flow and Metabolism.* 1996; 16: 1271–9.

[19] Baglivo J; Olivier D & Pagano M. Methods for the analysis of contingency tables with large and small cell counts. *JASA.* 1988; 83: 1006–13.

[20] Baglivo J; Olivier D & Pagano M. Methods for exact goodness-of-fit tests. *JASA.* 1992; 87: 464–469.

[21] Baker RD. Two permutation tests of equality of variance. *Stat. & Comput.* 1995; 5(4): 289–96.

[22] Barbe P; Bertail P. *The Weighted Bootstrap.* New York: Springer-Verlag; 1995.

[23] Barbella P; Denby L & Glandwehr JM. Beyond exploratory data analysis: The randomization test. *Math Teacher.* 1990; 83: 144–49.

[24] Barton DE; David FN. Randomization basis for multivariate tests. *Bull. Int. Statist. Inst.* 1961; 39(2): 455–67.

[25] Basu D. Discussion of Joseph Berkson's paper "In dispraise of the exact test". *J. Statist. Plan. Infer.* 1979; 3: 189–92.

[26] Berger RL; Boos DD. P values maximized over a confidence set for a nuisance parameter. *JASA.* 1994; 89: 1012–16.

[27] Berger JO; Wolpert RW. *The Likelihood Principle.* IMS Lecture Notes—Monograph Series. Hayward CA:IMS; 1984.

[28] Berkson J. In dispraise of the exact test. *J. Statist. Plan. Inf.* 1978; 2: 27–42.

[29] Berry KJ; Kvamme KL & Mielke PWjr. Permutation techniques for the spatial analysis of the distribution of artifacts into classes. *Amer. Antiquity.* 1980; 45: 55–59.

[30] Berry KJ; Kvamme KL & Mielke PWjr. Improvements in the permutation test for the spatial analysis of the distribution of artifacts into classes. *Amer. Antiquity.* 1983; 48: 547–553.

[31] Berry KJ; Mielke PW Jr. A family of multivariate measures of association for nominal independent variables. *Educ.& Psych. Meas.* 1992; 52: 97–101.

[32] Berry KJ; Mileke PW Jr. A measure of association for nominal independent variables. *Educ.& Psych. Meas.* 1992; 52: 895–8.

[33] Besag JE. Some methods of statistical analysis for spatial data. *Bull. Int. Statist. Inst.* 1978; 47: 77–92.

[34] Bickel PM; Van Zwet WR. Asymptotic expansion for the power of distribution free tests in the two-sample problem. *Ann. Statist.* 1978; 6: 987–1004 (corr 1170–71).

[35] Bickel PJ; Freedman DA. Some asymptotic theory for the bootstrap. *Ann. Statist.* 1981; 9: 1196–1217.

[36] Bishop YMM; Fienberg SE & Holland PW. *Discrete Multivariate Analysis: Theory and Practice.* Cambridge MA: MIT Press; 1975.

[37] Blair C; Higgins JJ; Karinsky W; Krom R; & Rey JD. A study of multivariate permutation tests which may replace Hotellings T^2 test in prescribed circumstances. *Multivariate Beh.Res.* 1994; 29: 141–63.

[38] Blair RC; Troendle JF; & Beck, R. W. Control of familywise errors in multiple endpoint assessments via stepwise permutation tests. *Statistics in Medicine.* 1996; 15: 1107–1121

[39] Boess FG; Balasuvramanian R, Brammer MJ; & Campbell IC. Stimulation of mus-
      carinic acetylcholine receptors increases synaptosomal free calcium concentration by
      protein kinase-dependent opening of L-type calcium channels. *J. Neurochem.* 1990;
      55: 230–6.

[40] Boyett JM; Shuster JJ. Nonparametric one-sided tests in multivariate analysis with
      medical applications. *JASA.* 1977; 72: 665–668.

[41] Bradbury IS. Analysis of variance vs randomization tests: a comparison (with
      discussion by White and Still). *Brit. J. Math. Stat. Psych.* 1987; 40: 177–195.

[42] Bradley JV. *Distribution Free Statistical Tests.* New Jersey: Prentice Hall; 1968.

[43] Breiman L. The little bootstrap and other methods for density selection in regression:
      X-fixed prediction error. *JASA.* 1992; 87: 738–54.

[44] Breiman L; Friedman JH, Olshen RA; & Stone CJ. *Classification and Regression
      Trees.* Belmont CA: Wadsworth; 1984.

[45] Bross, IDJ. Taking a covariable into account. *JASA.* 1964; 59: 725–36.

[46] Bryant EH. Morphometric adaptation of the housefly, Musa domestica L; in the
      United States. *Evolution.* 1977; 31: 580–96.

[47] Buckland ST; Garthwaite PH. Quantifying precision of mark-recapture estimates
      using the bootstrap and related methods. *Biometrics.* 1991; 47: 225–68.

[48] Bullmore E; Brammer M; Williams SC; Rabehesk S; Janot N; David A; Mekers J;
      Howard R & Slam P. Statistical methods for estimation and inference for functional
      MR image analysis. *Magn. Res M.* 1996; 35(2): 261–77.

[49] Buonaccorsi JP. A note on confidence intervals for proportions in finite populations.
      Amer. Stat. 1987; 41: 215–8.

[50] Busby DG. Effects of aerial spraying of fenithrothion on breeding white-throated
      sparrows. *J. Appl. Ecol.* 1990; 27: 745–55.

[51] Cade B. Comparison of tree basal area and canopy cover in habitat models: Subalpine
      forest. *J. Alpine Mgmt.* 1997; 61: 326–35.

[52] Cade B; Hoffman H. Differential migration of blue grouse in Colorado. *Auk.* 1993;
      110:70–7.

[53] Cade B; Richards L. Permutation tests for least absolute deviation regression.
      Biometrics. 1996; 52: 886–902.

[54] Carter C; Catlett J. Assessing credit card applications using machine learning. *IEEE
      Expert.* 1987; 2: 71–79.

[55] Chhikara RK. State of the art in credit evaluation. *Amer. J. Agric. Econ.* 1989; 71:
      1138–44.

[56] Cleveland WS. *The Elements of Graphing Data.* Monterey CA: Wadsworth; 1985.

[57] Cleveland WS. *Visualizing Data.* Summit NJ: Hobart Press; 1993.

[58] Cliff AD; Ord JK. Evaluating the percentage points of a spatial autocorrelation
      coefficient. *Geog. Anal.* 1971; 3: 51-62

[59] Cliff AD; Ord JK. *Spatial Processes: Models and Applications.* London: Pion Ltd;
      1981.

[60] Cox DF; Kempthorne O. Randomization tests for comparing survival curves.
      Biometrics. 1963; 19: 307–17.

[61] Cox DR. Partial likelihood. *Biometrika.* 1975; 62: 269–76.

[62] Cox DR. Regression models and life tables (with discussion). *J. Roy. Statist. Soc B.*
      1972; 34: 187–220.

[63] Cox DR; Snell EJ. *Applied Statistics. Principles and Examples.* London: Chapman
      and Hall; 1981.

[64] Cover T; Hart P. Nearest neighbor pattern classification. *IEEE Trans. on Info. Theory.*
      1967; IT-13: 21–27.

[65] Dalen J. Computing elementary aggregates in the Swedish Consumer Price Index. *J. Official Statist.* 1992; 8: 129–147.

[66] Dansie BR. A note on permutation probabilities. *J. Roy. Statist. Soc. B.* 1983; 45: 22–24.

[67] Davis AW. On the effects of moderate nonnormality on Roy's largest root test. *JASA.* 1982; 77: 896-900.

[68] DeJager OC; Swanepoel JWH; & Raubenheimer BC. Kernel density estimators applied to gamma ray light curves. Astron. & Astrophy. 1986; 170: 187–96.

[69] Devroye L; Gyorfi L. *Nonparametric Density Estimation: The L1 View.* New York: Wiley; 1985.

[70] Diaconis P. Statistical problems in ESP research. *Science.* 1978; 201: 131-6.

[71] Diaconis P; Efron B. Computer intensive methods in statistics. *Sci. Amer.* 1983; 48: 116–130.

[72] DiCiccio TJ; Hall P; & Romano JP. On smoothing the bootstrap. *Ann. Statist.* 1989; 17: 692–704.

[73] DiCiccio TJ; Romano J. A review of bootstrap confidence intervals (with discussions). *J. Roy. Statist. Soc. B.* 1988; 50: 163–70.

[74] Dietz EJ. Permutation tests for the association between two distance matricies. *Systemic Zoology.* 1983; 32: 21–26.

[75] Diggle PJ; Lange N; & Benes FM. Analysis of variance for replicated spatial point paterns in Clinical Neuroanatomy. *JASA.* 1991; 86: 618–25.

[76] Do KA; Hall P. On importance resampling for the bootstrap. *Biometrika.* 1991; 78: 161–7.

[77] Doolittle RF. Similar amino acid sequences: chance or common ancestory. *Science.* 1981; 214: 149–159.

[78] Douglas ME; Endler JA. Quantitative matrix comparisons in ecological and evolutionary investigations. *J. theoret. Biol.* 1982; 99: 777-95

[79] Draper D; Hodges JS; Mallows CL; & Pregibon D. Exchangeability and data analysis (with discussion). *J. Roy. Statist. Soc. A.* 1993; 156: 9–28.

[80] Dubuisson B; Lavison P. Surveillance of a nuclear reactor by use of a pattern recognition methodology. *IEEE Trans. Systems, Man, & Cybernetics.* 1980; 10: 603–9.

[81] Ducharme GR; Jhun M; Romano JP; & Truong KN. Bootstrap confidence cones for directional data. *Biometrika.* 1985; 72: 637–45.

[82] Duffy DE; Quinoz AJ. Permutation-based algorithm for block clustering. *J. Classification.* 1991; 8: 65–91.

[83] Duffy DE; Fowlkes EB; & Kane LD. Cluster analysis in strategic data architecture design. *Bellcore Database Symposium*; 1987; Morristown NJ. Morristown NJ: Bell Communications Research, Inc.; 1987: 175–186.

[84] Dwass M. Modified randomization tests for non-parametric hypotheses. *Ann. Math. Statist.* 1957; 28: 181-7.

[85] Edgington ES. *Randomization Tests.* 3rd ed. New York: Marcel Dekker; 1995.

[86] Edgington ES. Validity of randomization tests for one-subject experiments. *J Educ Stat.* 1980; 5: 235–51.

[87] Efron B. Bootstrap methods: Another look at the jackknife. *Annals Statist.* 1979; 7: 1–26.

[88] Efron B. Censored data and the bootstrap. *JASA.* 1981; 76: 312–319.

[89] Efron, B. *The Jackknife, the Bootstrap and Other Resampling Plans.* Philadelphia: SIAM; 1982.

[90]   Efron B. Estimating the error rate of a prediction rule: improvements on cross-validation. *JASA*. 1983; 78: 316–31.

[91]   Efron B. Bootstrap confidence intervals: good or bad? (with discussion). *Psychol. Bull.* 1988; 104: 293–6.

[92]   Efron B. Six questions raised by the bootstrap. R. LePage and L. Billard, editors. *Exploring the Limits of the Bootstrap*. New York: Wiley; 1992.

[93]   Efron B. Missing data, imputation, and the bootstrap (with discussion). *JASA*. 1994; 89: 463–79.

[94]   Efron B; DiCiccio T. More accurate confidence intervals in exponential families. *Biometrika*. 1992; 79: 231–45.

[95]   Efron B; Tibshirani R. Bootstrap measures for standard errors, confidence intervals, and other measures of statistical accuracy. *Stat Sci.* 1986; 1: 54–77.

[96]   Efron B; Tibshriani R. Statistical data analysis in the computer age. *Science.* 1991; 253: 390–5.

[97]   Efron B; Tibshirani R. *An Introduction to the Bootstrap*. New York: Chapman and Hall; 1993.

[98]   Entsuah AR. Randomization procedures for analyzing clinical trend data with treatment related withdrawls. *Commun. Statist. A.* 1990; 19: 3859–80.

[99]   Falk M; Reiss RD. Weak convergence of smoothed and nonsmoothed bootstrap quantiles estimates. *Ann. Prob.* 1989; 17: 362–71.

[100] Faris PD; Sainsbury RS. The role of the Pontis Oralis in the generation of RSA activity in the hippocampus of the guinea pig. *Psych.& Beh.* 1990; 47: 1193–9.

[101] Farrar DA; Crump KS. Exact statistical tests for any cocarcinogenic effect in animal assays. *Fund. Appl. Toxicol.* 1988; 11: 652–63.

[102] Farrar DA; Crump KS. Exact statistical tests for any cocarcinogenic effect in animal assays. II age adjusted tests. *Fund. Appl. Toxicol.* 1991; 15: 710–21.

[103] Fears TR; Tarone RE; & Chu KC. False-positive and false-negative rates for carcinogenicity screens. *Cancer Res.* 1977; 37: 1941–5.

[104] Feinstein AR. Clinical biostatistics XXIII. The role of randomization in sampling, testing, allocation, and credulous idolatry (part 2). *Clinical Pharm.* 1973; 14: 989–1019.

[105] Festinger LC; Carlsmith JM. Cognitive consequences of forced compliance. *J. Abnorm. Soc. Psych.* 1959; 58: 203–10.

[106] Fisher RA. *Statistical Methods for Research Workers*. Edinburgh: Oliver & Boyd; 1st ed 1925.

[107] Fisher RA. The logic of inductive inference (with discussion). *J. Roy. Statist. Soc. A.* 1934; 98: 39–54.

[108] Fisher RA. *Design of Experiments*. New York: Hafner; 1935.

[109] Fisher RA. Coefficient of racial likeness and the future of craniometry. *J Royal Anthrop Soc.* 1936; 66: 57–63.

[110] Fisher RA. The use of multiple measurements in taxonomic problems. *Ann. Eugenics.* 1936; 7: 179-88.

[111] Fix E; Hodges JL Jr; & Lehmann EL. The restricted chi-square test. *Studies in Probability and Statistics Dedicated to Harold Cramer.* Stockholm: Almquist and Wiksell; 1959.

[112] Fleming TR; Harrington DP. *Counting Processes and Survival Analysis.* New York: John Wiley & Sons; 1991.

[113] Ford RD; Colom LV; & Bland BH. The classification of medial septum-diagonal band cells as theta-on or theta-off in relation to hippo campal EEG states. *Brain Res.* 1989; 493: 269–282.

[114] Forster JJ; McDonald, JW; & Smith, PWF. Monte Carlo exact conditional tests for log-linear and logistic models. *J. Roy. Statist. Soc. B.* 1966; 58: 445-453.

[115] Forsythe AB; Engleman L; Jennrich R. A stopping rule for variable selection in multivariate regression. *JASA.* 1973; 68: 75–7.

[116] Forsythe AB; Frey HS. Tests of significance from survival data. *Comp. & Biomed. Res.* 1970; 3: 124–132.

[117] Foutz RN; Jensen DR; & Anderson GW. Multiple comparisons in the randomization analysis of designed experiments with growth curve responses. *Biometrics.* 1985; 41: 29–37.

[118] Frank, D; Trzos RJ; & Good P. Evaluating drug-induced chromosome alterations. *Mutation Res.* 1978; 56: 311–7.

[119] Fraumeni JF; Li FP. Hodgkin's disease in childhood: an epidemiological study. *J. Nat. Cancer Inst.* 1969; 42: 681–91.

[120] Freedman D; Diaconis P. On the histogram as a density estimator: L2 theory. *Zeitschrift für Wahrscheinlichkeitstheorie und verwandte Gebeite.* 1981; 57: 453–76.

[121] Freedman L. Using permutation tests and bootstrap confidence limits to analyze repeated events data. *Contr. Clin. Trials.* 1989; 10: 129–141.

[122] Freeman GH; Halton JH. Note on an exact treatment of contingency, goodness of fit, and other problems of significance. *Biometrika.* 1951; 38: 141–149.

[123] Friedman JH. Exploratory projection pursuit. *JASA.* 1987; 82: 249–66.

[124] Fryer MJ. A review of some nonparametric methods of density estimation. *J. Instr. Math. Applic.* 1976; 20: 335–54.

[125] Fukunaga K. *Statistical Pattern Recognition* (2nd ed). London: Academic Press; 1990.

[126] Gabriel KR. Some statistical issues in weather experimentation. *Commun Statist. A.* 1979; 8: 975–1015.

[127] Gabriel KR; Hall, WJ. Rerandomization inference on regression and shift effects: Computationally feasible methods. *JASA.* 1983; 78: 827–36.

[128] Gabriel KR; Hsu CF. Evaluation of the power of rerandomization tests, with application to weather modification experiments. *JASA.* 1983; 78: 766–75.

[129] Gabriel KR; Sokal RR. A new statistical approach to geographical variation analysis. *Systematic Zoology.* 1969; 18: 259–70.

[130] Gail M; Mantel N. Counting the number of rxc contingency tables with fixed marginals. *JASA.* 1977; 72: 859–62.

[131] Gail MH; Tan WY; & Piantadosi S. Tests for no treatment effect in randomized clinical trials. *Biometrika.* 1988; 75: 57–64.

[132] Gart JJ. *Statistical Methods in Cancer Res.*, Vol III - The design and analysis of long term animal experiments. Lyon: IARC Scientific Publications; 1986.

[133] Garthwaite PH. Confidence intervals from randomization tests. *Biometrics.* 1996; 52: 1387–93.

[134] Gastwirht JL. Statistical reasoning in the legal setting. *Amer. Statist.* 1992; 46: 55–69.

[135] Geisser S. The predictive sample reuse method with applications. *JASA.* 1975; 70: 320-8.

[136] Gine E; Zinn J. Necessary conditions for a bootstrap of the mean. *Ann. Statist.* 1989; 17:684–91.

[137] Glass AG; Mantel N. Lack of time-space clustering of childhood leukemia, Los Angeles County 1960–64. *Cancer Res.* 1969; 29: 1995–2001.

[138] Glass AG; Mantel N; Gunz FW; & Spears GFS. Time-space clustering of childhood leukemia in New Zealand. *J. Nat. Cancer Inst.* 1971; 47: 329–36.

[139] Gleason JR. Algorithms for balanced bootstrap simulations. *Amer. Statistician.* 1988; 42: 263–66.

[140] Gliddentracey C; Greenwood, AK. A validation study of the Spanish self directed search using back translation procedures. *J. Career Assess.* 1997; 5: 105–13.

[141] Gliddentracey, CE; Parraga, MI. Assessing the structure of vocational interests among Bolivian university students. *J. Vocational Beh.* 1996; 48: 96–106.

[142] Goldberg P; Leffert F; Gonzales M; Gorgenola I; & Zerbe GO. Intraveneous aminophylline in asthma: A comparison of two methods of administration in children. *Amer. J. of Diseases and Children.* 1980; 134: 12–18.

[143] Good IJ; Gaskins RA. Density estimation and bump-hunting by the penalized likelihood method exemplified by scattering and meteorite data (with discussion). *JASA.* 1980; 75: 42–73.

[144] Good PI. Detection of a treatment effect when not all experimental subjects respond to treatment. *Biometrics.* 1979; 35(2): 483–9.

[145] Good PI. Almost most powerful tests for composite alternatives. *Comm. Statist.— Theory & Methods.* 1989; 18(5): 1913–25.

[146] Good PI. Most powerful tests for use in matched pair experiments when data may be censored. *J. Statist. Comp. Simul.* 1991; 38: 57–63.

[147] Good PI. Globally almost powerful tests for censored data. *Nonparametric Statistics.* 1992; 1: 253–62.

[148] Good PI. *Permutation Tests.* New York: Springer Verlag; 1994.

[149] Goodman L; Kruskal W. Measures of association for cross-classification. *JASA.* 1954; 49: 732–64.

[150] Gordon L; Olshen RA. Tree-structured survival analysis (with discussion). *Cancer Treatment Reports.* 1985; 69: 1065–8.

[151] Graubard BI; Korn EL. Choice of column scores for testing independence in ordered 2 by K contingency tables. *Biometrics.* 1987; 43: 471–6.

[152] Graves GW; Whinston AB. An algorithm for the quadratic assignment probability. *Mgmt. Science.* 1970; 17: 453-71.

[153] Grubbs G. Fiducial bounds on reliability for the two-parameter negative exponential distribution. *Technometrics.* 1971; 13: 873–6.

[154] Gupta et al. *Community Dentistry and Oral Epidemiology.* 1980; 8: 287–333.

[155] Haber M. A comparison of some conditional and unconditional exact tests for 2x2 contingency tables. *Comm. Statist. A.* 1987; 18: 147–56.

[156] Hall P. On the bootstrap and confidence intervals. *Ann. Statist.* 1986; 14: 1431–52.

[157] Hall P. Theoretical comparison of bootstrap confidence intervals (with discussion). *Ann. Statist.* 1988; 16: 927–85.

[158] Hall P; Martin MA. On bootstrap resampling and iteration. *Biometrika.* 1988; 75: 661–71.

[159] Hall P; Titterington M. The effect of simulation order on level accuracy and power of Monte Carlo tests. *J. Roy. Statist. Soc. B.* 1989; 51: 459–67.

[160] Hall P; Wilson SR. Two guidelines for bootstrap hypothesis testing. *Biometrics.* 1991; 47: 757–62.

[161] Hartigan JA. Using subsample values as typical values. *JASA.* 1969; 64: 1303–17.

[162] Hartigan JA. Error analysis by replaced samples. *J. Roy. Statist. Soc B.* 1971; 33: 98–110.

[163] Hartigan JA. Direct clustering of a data matrix. *JASA.* 1972; 67: 123–9.

[164] Hartigan JA. *Clustering Algorithms.* New York: John Wiley & Sons; 1975.

[165] Hårdle W. *Smoothing Techniques with Implementation in S.* New York: Springer-Verlag; 1991.

[166] Hårdle W; Bowman A. Bootstrapping in nonparametric regression: local adaptive smoothing and confidence bands. *JASA*. 1988; 83: 102–10.

[167] Hardy FL;Youse BK. *Finite Mathematics for the Managerial, Social, and Life Sciences*. New York: West Publishing; 1984.

[168] Hasegawa M; Kishino H; & Yano T. Phylogenetic inference from DNA sequence data. K. Matusita, editor. *Statistical Theory and Data Analysis*. Amsterdam: North Holland; 1988.

[169] Henery RJ. Permutation probabilities for gamma random variables. *J Appl Probab.* 1983; 20(4): 822–34.

[170] Henze N. A multivariate two-sample test based on the number of nearest neighbor coincidence. *Annals Statist.* 1988; 16: 772–83.

[171] Hettmansperger TP. *Statistical Inference Based on Ranks*. New York: Wiley, 1984.

[172] Higgins JJ; Noble W. A permutation test for a repeated measures design. *Applied Statistics in Agriculture.* 1993; 5: 240–55.

[173] Highton R. Comparison of microgeographic variation in morphological and electrophoretic traits. In Hecht MK, Steer WC, & B Wallace eds. *Evolutionary Biology.* New York: Plenum; 1977; 10: 397–436

[174] Hill AB. *Principles of Medical Statistics* (8th ed). London: Oxford University Press; 1966.

[175] Hinkley DV; Shi S. Importance sampling and the nested bootstrap. *Biometrika.* 1989; 76: 435–46.

[176] Hirji KF; Mehta CR; & Patel NR. Computing distributions for exact logistic regression. *JASA.* 1987; 82: 1110–7.

[177] Hjorth, JSU. *Computer Intensive Statistical Methods: Validation, Model Selection and Bootstrap.* New York: Chapman and Hall; 1994.

[178] Hodges JL; Lehmann EL. Estimates of location based on rank tests. *Ann. Math. Statist.* 1963; 34: 598–611.

[179] Hoeffding W. The large-sample power of tests based on permutations of observations. *Ann. Math. Statist.* 1952; 23: 169–92.

[180] Hoel DG; Walburg HE. Statistical analysis of survival experiments. *J. Nat. Cancer Inst.* 1972; 49: 361-72.

[181] Hollander M; Pena E. Nonparametric tests under restricted treatment assigment rules. *JASA.* 1988; 83(404): 1144–51.

[182] Hollander M; Sethuraman J. Testing for agreement between two groups of judges. *Biometrika.* 1978; 65: 403–12.

[183] Holmes MC; Williams REO. The distribution of carriers of streptococcus pyrogenes among 2413 healthy children. *J. Hyg. Camd.* 1954; 52: 165–79.

[184] Hosmer DW; Lemeshow S. Best subsets logistic regression. *Biometrics.* 1989.

[185] Hope ACA. A modified Monte Carlo significance test procedure. *J. Roy. Statist. Soc. B.* 1968; 30: 582–98.

[186] Howard M (pseud for Good P). Randomization in the analysis of experiments and clinical trials. *American Laboratory.* 1981; 13: 98–102.

[187] Hubert LJ. Combinatorial data analysis: Association and partial association. *Psychometrika.* 1985; 50: 449–67.

[188] Hubert LJ; Baker FB. Analyzing distinctive features confusion matrix. *J. Educ. Statist.* 1977; 2: 79–98.

[189] Hubert LJ; Baker FB. Evaluating the conformity of sociometric measurements. *Psychometrika.* 1978; 43: 31–42.

[190] Hubert LJ; Golledge RG; & Costanzo CM. Analysis of variance procedures based on a proximity measure between subjects. *Psych. Bull.* 1982; 91: 424–30.

[191] Hubert LJ; Golledge RG; Costanzo CM; Gale N; &. Halperin WC. Nonparametric tests for directional data. In Bahrenberg G, Fischer M. &. P. Nijkamp, eds. *Recent developments in spatial analysis: Methodology, Measurement, Models.* Aldershot UK: Gower; 1984: 171–90.

[192] Hubert LJ; Schultz J. Quadratic assignment as a general data analysis strategy. *Brit. J. Math. Stat. Psych.* 1976; 29: 190–241.

[193] Ingenbleek JF. Tests simultanes de permutation des rangs pour bruit-blanc multivarie. *Statist. Anal Donnees.* 1981; 6: 60–5.

[194] Izenman AJ. Recent developments in nonparametric density estimation. *JASA.* 1991; 86(413): 205–224.

[195] Izenman AJ.; Sommer CJ. Philatelic mixtures: multimodal densities. *JASA.* 1988; 83: 941–53.

[196] Jackson DA. Ratios in acquatic sciences: Statistical shortcomings with mean depth and the morphoedaphic index. *Canadian J Fisheries and Acquatic Sciences.* 1990; 47: 1788–95.

[197] Janssen A. Conditional rank tests for randomly censored data. *Annal Statist.* 1991; 19: 1434–56.

[198] Jennrich RI. A note on the behaviour of the log rank permutation test under unequal censoring. *Biometrika.* 1983; 70: 133–7.

[199] Jennrich RI. Some exact tests for comparing survival curves in the presence of unequal right censoring. *Biometrika.* 1984; 71: 57–64.

[200] Jin MZ. On the multisample pemutation test when the experimental units are nonuniform and random experimental errors exist. *J System Sci Math Sci.* 1984; 4: 117–27, 236–43.

[201] Jockel KH. Finite sample properties and asymptotic efficiency of Monte Carlo tests. *Annals Statistics.* 1986; 14: 336–347.

[202] John RD; Robinson J. Significance levels and confidence intervals for randomization tests. *J. Statist. Comput. Simul.* 1983; 16: 161–73.

[203] Johns MV jr. Importance sampling for bootstrap confidence intervals. *JASA.* 1988; 83: 709–14.

[204] Johnstone IM.; Silverman BW. Speed of estimate in positron emission tomography and related studies. *Annal Statist.* 1990; 18: 251–80.

[205] Jones HL. Investigating the properties of a sample mean by employing random subsample means. *JASA.* 1956; 51: 54–83.

[206] Jorde LB; Rogers AR; Bamshad M; Watkins WS; Krakowiak P; Sung S; Kere J; & Harpending HC. Microsatellite diversity and the demographic history of modern humans. *Proc. Nat. Acad. Sci.* 1997; 94: 3100–3.

[207] Kalbfleisch JD. Likelihood methods and nonparametric tests. *JASA.* . 1978; 73: 167–170.

[208] Kalbfleisch JD; Prentice RL. *The Statistical Analysis of Failure Time Data.* New York: John Wiley & Sons; 1980.

[209] Kaplan EL; Meier P. Non-parametric estimation from incomplete observations. *JASA.* 1958; 53: 457–81, 562–3.

[210] Karlin S; Ghandour G; Ost F; Tauare S; & Korph K. New approaches for computer analysis of DNA sequences. *Proc. Nat. Acad. Sci., USA.* 1983; 80: 5660–4.

[211] Karlin S; Williams PT. Permutation methods for the structured exploratory data analysis (SEDA) of familial trait values. *Amer. J. Human Genetics.* 1984; 36: 873–98.

[212] Kazdin AE. Statistical analysis for single-case experimental designs. In *Strategies for Studying Behavioral Change.* M Hersen and DH Barlow, editors. New York: Pergammon; 1976.

[213] Kazdin, AE. Obstacles in using randomization tests in single-case experiments. *J Educ Statist.* 1980; 5: 253–60.

[214] Keller-McNulty S.; Higgens JJ. Effect of tail weight and outliers on power and type I error of robust permuatation tests for location. *Commun. Stat.—Theory and Methods.* 1987; 16: 17–35.

[215] Kempthorne O. The randomization theory of experimental inference. *JASA.* 1955; 50: 946–67.

[216] Kempthorne O. Some aspects of experimental inference. *JASA.* 1966; 61: 11–34.

[217] Kempthorne O. Inference from experiments and randomization. In *A Survey of Statistical Design and Linear Models.* JN Srivastava. Amsterdam: North Holland; 1975: 303–32.

[218] Kempthorne O. Why randomize? *J. Statist. Prob. Infer.* 1977; 1: 1–26.

[219] Kempthorne O. In dispraise of the exact test: reactions. *J. Statist. Plan. Infer.* 1979; 3: 199–213.

[220] Kempthorne O; Doerfler TE. The behavior of some significance tests under experimental randomization. *Biometrika.* 1969; 56: 231–48.

[221] Klauber MR. Two-sample randomization tests for space-time clustering. *Biometrics.* 1971; 27: 129–42.

[222] Klauber MR; Mustacchi A. Space-time clustering of childhood leukemia in San Francisco. *Cancer Res.* 1970; 30: 1969–73.

[223] Kleinbaum DB; Kupper LL; & Chambless LE. Logistic regression analysis of epidemiologic data: theory and practice. *Commun. Statist. A.* 1982; 11: 485–547.

[224] Knight K. On the bootstrap of the sample mean in the infinite variance case. *Annal Statist.* 1989; 17: 1168–73.

[225] Koch G (ed). *Exchangeability in Probability and Statistics.* Amsterdam: North Holland; 1982.

[226] Koziol JA; Maxwell DA; Fukushima M; Colmer A; & Pilch YHA distribution-free test for tumor-growth curve analyses with applications to an animal tumor immunotherapy experiment. *Biometrics.* 1981; 37: 383–90.

[227] Krehbiel K. Are Congressional committees composed of preference outliers? *Amer. Poli. Sci. Rev.* 1990; 84: 149–63.

[228] Krewski D; Brennan J; & M Bickis. The power of the Fisher permutation test in 2 by k tables. *Commun. Stat. B.* 1984; 13: 433–48.

[229] Kryscio RJ; Meyers MH; Prusiner SI; Heise HW; & BW Christine. The space-time distribution of Hodgkin's disease in Connecticut, 1940-1969. *J. Nat. Cancer Inst.* 1973; 50: 1107–10.

[230] Lachin JM. Properties of sample randomization in clinical trials. *Contr. Clin. Trials.* 1988; 9: 312–26.

[231] Lachin JN. Statistical properties of randomization in clinical trials. *Contr. Clin. Trials.* 1988; 9: 289–311.

[232] Lahri SN. Bootstrapping the studentized sample mean of lattice variables. *J. Multiv. Anal.* 1993; 45:247–56.

[233] Lambert D. Robust two-sample permutation tests. *Ann. Statist.* 1985; 13: 606–25.

[234] Leemis LM. Relationships among common univariate distributions. *Amer. Statistician.* 1986; 40: 143–6.

[235] Lefebvre M. Une application des methodes sequentielles aux tests de permutations. *Canad. J. Statist.* 1982; 10: 173–80.

[236] Lehmann EL. *Testing Statistical Hypotheses.* New York: John Wiley & Sons; 1986.

[237] Leonard T. Density estimation, stochastic processes, and p-information (with discussion). *J. Roy. Statist. Soc. B.* 1978; 40: 113–46.

[238] Levin DA. The organization of genetic variability in Phlox drummondi. *Evolution.* 1977; 31: 477–94.

[239] Liu RY. Bootstrap procedures under some non iid models. *Ann. Statist.* 1988; 16: 1696–1788.

[240] Livezey RE. Statistical analysis of general circulation model climate simulation: Sensitivity and prediction experiments. *J Atmospheric Sciences.* 1985; 42: 1139–49.

[241] Livezey RE; Chen W. Statistical field significance and its determination by Monte Carlo techniques. *Monthly Weather Review.* 1983; 111: 46–59.

[242] Loh WY. Estimating an endpoint of a distribution with resampling methods. *Ann. Statist.* 1984; 12: 1543–50.

[243] Loh WY. Bootstrap calibration for confidence interval construction and selection. *Statist. Sinica.* 1991; 1: 479–95.

[244] Loughin TM; Noble W. A permutation test for effects in an unreplicated factorial design. *Technometrics.* 1997; 39: 180–90.

[245] Lunneborg CE. Estimating the correlation coefficient: The bootstrap approach. *Psychol. Bull.* 1985; 98: 209–15.

[246] Mackay DA; Jones RE. Leaf-shape and the host-finding behavior of two ovipositing monophagous butterfly species. *Ecol. Entom.* 1989; 14: 423–31.

[247] Macuson R; Nordbrock E. A multivariate permutation test for the analysis of arbitrarily censored survival data. *Biometrical J.* 1981; 23: 461–5.

[248] Makinodan T; Albright JW; Peter CP; Good PI; & Hedrick ML. Reduced humoral activity in long-lived mice. *Immunology.* 1976; 31: 400–8.

[249] Makuch RW; Parks WP. Response to serum antigen level to AZT for the treatment of AIDS. *AIDS Research and Human Retroviruses.* 1988; 4: 305–16.

[250] Manly BFJ. The comparison and scaling of student assessment marks in several subjects. *Applied Statistics.* 1988; 37: 385–95.

[251] Manly BFJ. *Randomization, Bootstrap and Monte Carlo Methods in Biology.* (2nd ed.). London: Chapman & Hall; 1997.

[252] Mann RC; Hand RE Jr. The randomization test applied to flow cytometric histograms. *Computer Programs in Biomedicine.* 1983; 17: 95–100.

[253] Mantel N. The detection of disease clustering and a generalized regression approach. *Cancer Res.* 1967; 27: 209–220.

[254] Mantel N; Bailar JC. A class of permutational and multinomial tests arising in epidemiological research. *Biometrics.* 1970; 26: 687–700.

[255] Mapleson WW. The use of GLIM and the bootstrap in assessing a clinical trial of two drugs. *Statist. Med.* 1986; 5: 363–74.

[256] Marcus LF. Measurement of selection using distance statistics in prehistoric orang-utan pongo pygamous palaeosumativens. *Evolution.* 1969; 23: 301.

[257] Mardia KV; Kent JT and Bibby JM. *Multivariate Analysis.* New York: Academic Press; 1979.

[258] Maritz JS. *Distribution Free Statistical Methods.* (2nd ed.) London: Chapman & Hall; 1996.

[259] Marriott FHC. Barnard's Monte Carlo tests: How many simulations? *Appl. Statist.* 1979; 28: 75–7.

[260] Marron JS. A comparison of cross-validation techniques in density estimation. *Ann. Statist.* 1987; 15: 152–62.

[261] Martin MA. On bootstrap iteration for coverage of confidence intervals. *JASA.* 1990; 85:1105–8.

[262] Martin-Lof P. Exact tests, confidence regions and estimates. In Barndorff-Nielsen O; Blasild P; & Schow G. *Proceeding of the Conference of Foundational Questions in*

Statistical Inference. Aarhus: Institute of Mathematics, University of Aarhus; 1974; 1: 121–38.

[263] Maxwell SE; Cole DA. A comparison of methods for increasing power in randomized between-subjects designs. *Psych Bull.* 1991; 110: 328–37.

[264] May RB; Masson MEJ; & Hunter MA. *Application of Statistics in Behavioral Research.* New York: Harper & Row; 1990.

[265] McCarthy PJ. Psuedo-replication: Half samples. *Review Int. Statist. Inst.* 1969; 37: 239–64.

[266] McKinney PW; Young MJ, Hartz A, Bi-Fong Lee M. The inexact use of Fisher's exact test in six major medical journals. *J. American Medical Association.* 1989; 261: 3430–3.

[267] Mehta CR. An interdisciplinary approach to exact inference for contingency tables. *Statist. Sci.* 1992; 7: 167–70.

[268] Mehta CR; Patel NR. A network algorithm for the exact treatment of the 2xK contingency table. *Commun. Statist. B.* 1980; 9: 649–64.

[269] Mehta CR; Patel NR. A network algorithm for performing Fisher's exact test in rxc contingency tables. *JASA.* 1983; 78: 427–34.

[270] Mehta CR; Patel NR and Gray R. On computing an exact confidence interval for the common odds ratio in several 2x2 contingency tables. *JASA.* . 1985; 80: 969–73.

[271] Mehta CR; Patel NR and Senchaudhuri P. Importance sampling for estimating exact probabilities in permutational inference. *JASA.* 1988; 83: 999–1005.

[272] Melia KF; Ehlers CL. Signal detection analysis of ethanol effects on a complex conditional discrimination. *Pharm Biochem Behavior.* 1933; 1989: 581–4.

[273] Merrington M; Spicer CC. Acute leukemia in New England. *Brit J Preventive and Social Medicine.* 1969; 23: 124–7.

[274] Micceri, T. The unicorn, the normal curve, and other improbable creatures. *Psychol. Bull.* 1989; 105: 156–66.

[275] Mielke PW. Some parametric, nonparametric and permutation inference procedures resulting from weather modification experiments. *Commun Statist. A.* 1979; 8: 1083–96.

[276] Mielke PW. Meterological applications of permutation techniques based on distance functions. In Krishnaiah PR; Sen PK, editors. *Handbook of Statistics.* Amsterdam: North-Holland; 1984; 4: 813–30.

[277] Mielke PW Jr. Geometric concerns pertaining to applications of statistical tests in the atmospheric sciences. *J. Atmospheric Sciences.* 1985; 42: 1209–12.

[278] Mielke PW. Non-metric statistical analysis: Some metric alternatives. *J. Statist. Plan. Infer.* 1986; 13: 377–87.

[279] Mielke PW Jr. The application of multivariate permutation methods based on distance functions in the earth sciences. *Earth-Science Rev.* 1991; 31: 55–71.

[280] Mielke PW; Berry KJ. An extended class of permutation techniques for matched pairs. *Commun. Statist.—Theory & Meth.* 1982; 11: 1197–1207.

[281] Mielke PW Jr; Berry KJ. Fisher's exact probability test for cross-classification tables. *Educational and Psychological Measurement.* 1992; 52: 97–101.

[282] Miller AJ; Shaw DE; Veitch LG; & Smith EJ. Analyzing the results of a cloud-seeding experiment in Tasmania. *Commun. Statist. A.* 1979; 8: 1017–47.

[283] Mitchell-Olds T. Analysis of local variation in plant size. *Ecology.* 1987; 68: 82–7.

[284] Mitchell-Olds T. Quantitative genetics of survival and growth in Impatiens capensis. *Evolution.* 1986; 40: 107–16.

[285] Mooney CZ; Duval RD. *Bootstrapping: A Nonparametric Approach to Statistical Inference.* Newbury Park CA: Sage Publications; 1993.

[286] Mueller LD; Altenberg L. Statistical inference on measures of niche overlap. *Ecology.* 1985; 66: 1204–10.

[287] Mukhopadhyay I. Nonparametric tests for multiple regression under permutation symmetry. Calcutta Statist. *Assoc. Bull.* 1989; 38: 93–114.

[288] Nelson W. *Applied Life Data Analysis.* New York: John Wiley & Sons; 1982.

[289] Neyman J. *First Course in Probability and Statistics.* New York: Holt; 1950.

[290] Neyman J; Scott E. Field galaxies: luminosity, redshift, and abundance of types. Part I: Theory. *Proceedings of the 4th Berkeley Symposium.* 1960; 3: 261–76.

[291] Nguyen TT. A generalization of Fisher's exact test in pxq contingency tables using more concordant relations. *Commun. Statist. B.* 1985; 14: 633–45.

[292] Noether, GE. Distribution-free confidence intervals. *Statistica Neerlandica.* 1978; 32: 104–122.

[293] Noreen E. *Computer Intensive Methods for Testing Hypotheses.* New York: John Wiley & Sons; 1989.

[294] Oden NL. Allocation of effort in Monte Carlo simulations for power of permutation tests. *JASA.* 1991; 86: 1074–76.

[295] Oja H. On permutation tests in multiple regression and analysis of covariance problems. *Australian J Statist.* 1987; 29: 91–100.

[296] Park BU; Marron JS. Comparison of data-driven bandwidth selectors. *JASA.* 1990; 86: 66–72.

[297] Parzan E. On the estimation of a probability density function and the mode. *Ann. Math. Statist.* 1961; 33: 1065–76.

[298] Passing H. Exact simultaneous comparisons with controls in an rxc contingency table. *Biometrical J.* 1984; 26: 643–54.

[299] Patefield WM. Exact tests for trends in ordered contingency tables. *Appl. Statist.* 1982; 31: 32–43.

[300] Patil CHK. Cochran's Q test: exact distribution. *JASA.* 1975; 70: 186–9.

[301] Pearson ES. Some aspects of the problem of randomization. *Biometrika.* 1937; 29: 53–64.

[302] Peck R; Fisher L; & van Ness J. Bootstrap confidence intervals for the numbers of clusters in cluster analysis. *JASA.* 1989; 84: 184–91.

[303] Penninckx W; Hartmann C; Massart DL; & Smeyersverbeke J. Validation of the calibration procedure in atomic absorption spectrometric methods. *J Analytical Atomic Spectrometry.* 1996; 11: 237–46.

[304] Peritz E. Exact tests for matched pairs: studies with covariates. *Commun. Statist. A.* 1982; 11: 2157–67 (errata 12: 1209–10).

[305] Peritz E. Modified Mantel-Haenszel procedures for matched pairs. *Commun. Statist. A.* 1985; 14: 2263–85.

[306] Petrondas DA; Gabriel RK. Multiple comparisons by rerandomization tests. *JASA.* 1983; 78: 949–957.

[307] Pitman EJG. Significance tests which may be applied to samples from any population. *Roy. Statist. Soc. Suppl.* 1937; 4: 119–30, 225–32.

[308] Pitman EJG. Significance tests which may be applied to samples from any population. Part III. The analysis of variance test. *Biometrika.* 1938; 29: 322–35.

[309] Plackett RL. Random permutations. *J Roy Statist. Soc B.* 1968; 30: 517–534.

[310] Plackett RL. Analysis of permutations. *Appl. Statist.* 1975; 24: 163–71.

[311] Plackett RL; Hewlett PS. A unified theory of quantal responses to mixtures of drugs. The fitting to data of certain models for two non-interactive drugs with complete positive correlation of tolerances. *Biometrics.* 1963; 19: 517–31.

[312] Prager MH; Hoenig JM. Superposed epoch analysis: A randomization test of environmental effects on recruitment with application to chub mackrel. *Trans. Amer. Fisheries Soc.* 1989; 18: 608–19.

[313] Praska Rao BLS. *Nonparametric Functional Estimation.* New York: Academic Press; 1983.

[314] Priesendorfer RW; Barnett TP. Numerical model/reality intercomparison tests using small-sample statistics. *J Atmospheric Sciences.* 1983; 40: 1884–96.

[315] Rasmussen J. Estimating correlation coefficients: bootstrap and parametric approaches. *Psych. Bull.* 1987; 101: 136–9.

[316] Raz J. Testing for no effect when estimating a smooth function by nonparametric regression: a randomization approach. *JASA.* 1990; 85: 132–8.

[317] Richards LE; Byrd J. AS 304: Fisher's randomization test for two small independent samples. *Applied Statistics.* 1996; 45: 394–8.

[318] Ripley BD. Statistical aspects of neural networks. Barndorff-Nielsen OE, Jensen JL and Kendall WS, editors. *Networks and Chaos: Statistical and Probabilistic Aspects.* London: Chapman and Hall; 1993: 40–123.

[319] Roeder R. Density estimation with confidence sets exemplified by superclusters and voids in galaxies. *JASA.* 1990; 85: 617–24.

[320] Romano JP. A bootstrap revival of some nonparametric distance tests. *JASA.* 1988; 83: 698–708.

[321] Romano JP. On the behavior of randomization tests without a group invariance assumption. *JASA.* . 1990; 85(411): 686–92.

[322] Rosenbaum PR. Permutation tests for matched pairs with adjustments for covariates. *Appl. Statist.* 1988; 37(3): 401–11.

[323] Rosenblatt M. Remarks on some nonparametric estimators of a density function. *Ann. Math. Statist.* 1956; 27: 832–7.

[324] Royaltey HH; Astrachen E and Sokal RR. Tests for patterns in geographic variation. *Geographic Analysis.* 1975; 7: 369–95.

[325] Runger GC; Eaton MI. Most powerful invariant tests. *J. Multiv. Anal.* 1992; 42: 202–09.

[326] Ryan JM; Tracey TJG; & Rounds J. Generalizability of Holland's structure of vocational interests across ethnicity, gender, and socioeconomic status. *J. Counseling Psych.* 1996; 43: 330–7.

[327] Ryman N; Reuterwall C; Nygren K; & Nygren T. Genetic variation and differentiation in Scandiavian moose (Alces Alces): Are large mammals monomorphic? *Evolution.* 1980; 34: 1037–49.

[328] Saitoh T; Stenseth NC; & Bjornstad ON. Density dependence in fluctuating grey-sided vole populations. *J. Animal Ecol.* 1997; 66: 14–24.

[329] Salapatek P; Kessen W. Visual scanning of triangles by the human newborn. *J Exper Child Psych.* 1966; 3: 155–67.

[330] Scheffe H. *Analysis of Variance.* New York: John Wiley & Sons; 1959.

[331] Schemper M. A survey of permutation tests for censored survival data. *Commun Stat A.* 1984; 13: 433–48.

[332] Schenker N. Qualms about Bootstrap confidence intervals. *JASA.* 1985; 80: 360–1.

[333] Schultz JR; Hubert L. A nonparametric test for the correspondence between two proximity matrices. *J. Educ. Statist.* 1976; 1: 59–67.

[334] Scott DW. *Multivariate Density Estimation: Theory, Practice, and Visualization.* New York: John Wiley & Sons; 1992.

[335] Scott DW; Gotto AM; Cole JS: & Gorry JA. Plasma lipids as collateral risk factors in coronary artery disease—a study of 371 males with chest pain. *J. Chronic Diseases.* 1978; 31: 337–45.

[336] Scott DW; Thompson JR. Probability density estimation in higher dimensions. In Gentle JE ed. *Computer Science and Statistics: Proceedings of the Fifteenth Symposium on the Interface.* Amsterdam: North Holland; 1983: 173–9.

[337] Selander RK; Kaufman DW. Genetic structure of populations of the brown snail (Helix aspersa). I Microgeographic variation. *Evolution.* 1975; 29: 385–401.

[338] Shao J; Tu D. *The Jacknife and the Bootstrap.* New York: Springer; 1995.

[339] Shen CD; Quade D. A randomization test for a three-period three-treatment crossover experiment. *Commun. Statist. B.* 1986; 12: 183–99.

[340] Shuster JJ. *Practical Handbook of Sample Size Guidelines for Clinical Trials.* Boca Raton FL: CRC Press; 1993.

[341] Shuster JJ; Boyett JM. Nonparametric multiple comparison procedures. *JASA.* . 1979; 74: 379–82.

[342] Siemiatycki J. Mantel's space-time clustering statistic: computing higher moments and a comparison of various data transforms. *J Statist. Comput. Simul.* 1978; 7: 13–31.

[343] Siemiatycki J; McDonald AD. Neural tube defects in Quebec: A search for evidence of 'clustering' in time and space. *Brit. J. Prev. Soc. Med.* 1972; 26: 10–14.

[344] Silverman BW. Density ratios, empirical likelihood, and cot death. *Appl Statist.* 1978; 27: 26–33.

[345] Silverman BW. Using kernel density estimates to investigate multimodality. *J. Roy. Statist. Soc. B.* 1981; 43: 97–99.

[346] Silverman BW. *Density Estimation for Statistics and Data Analysis.* London: Chapman and Hall; 1986.

[347] Silverman BW; Young GA. The bootstrap: to smooth or not to smooth. *Biometrika.* 1987; 74: 469–79.

[348] Simon JL. *Basic Research Methods in Social Science.* New York: Random House; 1969.

[349] Singh K. On the asymptotic accuracy of Efron's bootstrap. *Ann. Statist.* 1981; 9: 1187–95.

[350] Smith DWF; Forester JJ; & McDonald JW. Monte Carlo exact tests for square contingency tables. *J Roy Statist. Soc A.* 1996; 159(2): 309–21.

[351] Smith PG; Pike MC. Generalization of two tests for the detection of household aggregation of disease. *Biometrics.* 1976; 32: 817–28.

[352] Smith PWF; McDonald JW; Forster JJ; & Berrington AM. Monte Carlo exact methods used for analysing inter-ethnic unions in Great Britain. *Appl. Statist.* 1996; 45: 191–202.

[353] Smythe RT. Conditional inference for restricted randomization designs. *Ann. Math. Statist.* 1988; 16(3): 1155–61.

[354] Sokal RR. Testing statistical significance in geographical variation patterns. *Systematic Zoo.* 1979; 28: 227–32.

[355] Solomon H. Confidence intervals in legal settings. In *Statistics and The Law.* DeGroot MH; Fienberg SE; & Kadane JB, ed. New York: John Wiley & Sons; 1986: 455–473.

[356] Solow AR. A randomization test for misclassification problems in discriminatory analysis. *Ecology.* 1990; 71: 2379–82.

[357] Soms AP. Permutation tests for k-sample binomial data with comparisons of exact and approximate P-levels. *Commun. Statist. A.* 1985; 14: 217–33.

[358] Stilson DW. *Psychology and Statistics in Psychological Research and Theory.* San Francisco: Holden Day; 1966.

[359] Stine R. An introduction to bootstrap methods: examples and ideas In *Modern Methods of Data Analysis.* J. Fox & J.S. Long, eds. Newbury Park CA: Sage Publications; 1990: 353–373.

[360] Stone M. Cross-validation choice and assessment of statistical predictions. *JASA.* 1974; B36: 111–147.

[361] Suissa S; Shuster JJ. Are uniformly most powerful unbiased tests really best? *Amer. Statistician.* 1984; 38: 204–6.

[362] Syrjala SE. A statistical test for a difference between the spatial distributions of two populations. *Ecology.* 1996; 77, 75–80.

[363] Tarter ME.; Lock MD. *Model-Free Curve Estimation.* New York: Chapman and Hall; 1993.

[364] Thompson JR.; Bridges E and Ensor K. Marketplace competition in the personal computer industry. *Decision Sciences.* 1992: 467–77.

[365] Thompson JR.; Tapia RA. *Nonparametric Function Estimation, Modeling and Simulation.* Philadelphia PA: SIAM; 1990.

[366] Thompson JR.; Scott D. Probability density estimation in higher dimensions. In J. Gentle ed. *Computer Science and Statistics.* Amsterdam: North Holland; 1983: 173–9.

[367] Thompson JR; Bartoszynski R; Brown B; & McBride C. Some nonparametric techniques for estimating the intensity function of a cancer related nonstationary Poisson process. *Ann. Statist.* 1981; 9: 1050–60.

[368] Tibshirani RJ. Variance stabilization and the bootstrap. *Biometrika.* 1988; 75: 433–44.

[369] Titterington DM; Murray GD; Spiegelhalter DJ; Skene AM; Habbema JDF; & Gelke GJ. Comparison of discrimination techniques applied to a complex data set of head-injured patients. *J Roy Statist. Soc A.* 1981; 144: 145–175.

[370] Tracy DS; Khan KA. Comparison of some MRPP and standard rank tests for three equal sized samples. *Commun. Statist. B.* 1990; 19: 315–33.

[371] Tracy DS; Tajuddin IH. Empirical power comparisons of two MRPP rank tests. *Commun. Statist. A.* 1986; 15: 551–70

[372] Tritchler D. On inverting permutation tests. *JASA.* 1984; 79: 200–207.

[373] Troendle JF. A stepwise resampling method of multiple hypothesis testing. *JASA.* 1995; 90: 370-8.

[374] Tsutakawa RK; Yang SL. Permutation tests applied to antibiotic drug resistance. *JASA.* 1974; 69: 87–92.

[375] Tukey JW. Improving crucial randomized experiments—especially in weather modification—by double randomization and rank combination. In LeCam L; & Binckly P, eds. *Proceeding of the Berkeley Conference in Honor of J Neyman and J Kiefer.* Heyward Ca: Wadsworth; 1985; 1: 79–108.

[376] Valdesperez RE. Some recent human-computer studies in science and what accounts for them. *AI Magazine.* 1995; 16: 37–44.

[377] van der Voet H. Comparing the predictive accuracy of models using a simple randomization test. *Chemometrics and Intelligent Laboratory Systems.* 1994; 25: 313–23.

[378] vanKeerberghen P; Vandenbosch C; Smeyers-Verbeke J; & Massart DL. Some robust statistical procedures applied to the analysis of chemical data. *Chemometrics and Intelligent Laboratory Systems.* 1991; 12: 3–13.

[379] van-Putten B. On the construction of multivariate permutation tests in the multivariate two-sample case. *Statist. Neerlandica.* 1987; 41: 191-201.

[380] Vecchia DF; HK Iyer. Exact distribution-free tests for equality of several linear models. *Commun Statist. A.* 1989; 18: 2467–88.

[381] Wald A; Wolfowitz J. An exact test for randomness in the nonparametric case based on serial correlation. *Ann. Math. Statist.* 1943; 14: 378–88.

[382] Wald A; Wolfowitz J. Statistical tests based on permutations of the observations. *Ann. Math. Statist.* 1944; 15: 358–72.

[383] Wåhrendorf J; Brown CC. Bootstrapping a basic inequality in the analysis of the joint action of two drugs. *Biometrics.* 1980; 36: 653–7.

[384] Wei LJ. Exact two-sample permutation tests based on the randomized play-the-winner rule. *Biometrika.* 1988; 75: 603–05.

[385] Wei LJ; Smythe RT; & Smith RL. K-treatment comparisons in clinical trials. *Annals Math Statist.* 1986; 14: 265–74.

[386] Welch WJ. Construction of permutation tests. *JASA.* 1990; 85(411): 693–8.

[387] Welch WJ. Rerandomizing the median in matched-pairs designs. *Biometrika.* 1987; 74: 609–14.

[388] Welch WJ; Guitierrez LG. Robust permutation tests for matched pairs designs. *JASA.* 1988; 83: 450–61.

[389] Werner M; Tolls R; Hultin J; & Mellecker J. Sex and age dependence of serum calcium, inorganic phosphorous, total protein, and albumin in a large ambulatory population. In *Fifth Technical International Congress of Automation, Advances in Automated Analysis.* Mount Kisco NY: Future Publishing; 1970.

[390] Wertz W. *Statistical Density Estimation: A Survey.* Gottingen: Vanerheck and Ruprecht; 1978.

[391] Westfall DH; Young SS. *Resampling-Based Multiple Testing: Examples and Methods for p-value Adjustment.* New York: John Wiley; 1993.

[392] Whaley FS. The equivalence of three individually derived permutation procedures for testing the homogenity of multidimensional samples. *Biometrics.* 1983; 39: 741–5.

[393] Whittaker J. *Graphical Models in Applied Statistics.* Chichester: Wiley; 1990.

[394] Wilk MB. The randomization analysis of a generalized randomized block design. *Biometrika.* 1955; 42: 70–79.

[395] Wilk MB; Kempthorne O. Nonadditivities in a Latin square design. *JASA.* 1957; 52: 218–36.

[396] Wilk MB; Kempthorne O. Some aspects of the analysis of factorial experiments in a completely randomized design. *Ann. Math. Statist.* 1956; 27: 950–84.

[397] Williams-Blangero S. Clan-structured migration and phenotypic differentiation in the Jirels of Nepal. *Human Biology.* 1989; 61: 143–57.

[398] Witztum D; Rips E; & Rosenberg Y. Equidistant letter sequences in the Book of Genesis. *Statist. Science.* 1994; 89: 768–76.

[399] Wong RKW; Chidambaram N; & Mielke PW. Applications of multi-response permutation procedures and median regression for covariate analyses of possible weather modification effects on hail responses. *Atmosphere-Ocean.* 1983; 21: 1–13.

[400] Yucesan E. Randomization tests for initialization bias in simulation output. *Naval Res. Logistics.* 1993; 40: 643–63.

[401] Young GA. Bootstrap: More than a stab in the dark. *Statist. Science.* 1994; 9: 382–415.

[402] Young GA. A note on bootstrapping the correlation coefficient. *Biometrika.* 1988; 75: 370–3.

[403] Zempo N; Kayama N; Kenagy RD; Lea HJ; & Clowes AW. Regulation of vascular smooth-muscle-cell migration and proliferation in vitro and in injured rat arteries by a synthetic matrix metalloprotinase inhibitor. *Art Throm V.* 1996; 16: 28–33.

[404] Zerbe GO. Randomization analysis of the completely randomized design extended to growth and response curves. *JASA*. 1979; 74: 215–21.

[405] Zerbe GO. Randomization analysis of randomized block design extended to growth and response curves. *Commun Statist. A*. 1979; 8: 191–205.

[406] Zerbe GO; Murphy JR. On multiple comparisons in the randomization analysis of growth and response curves. *Biometrics*. 1986; 42: 795–804.

[407] Zimmerman H. Exact calculations of permutation distributions for r dependent samples. *Biometrical J*. 1985; 27: 349–52.

[408] Zimmerman H. Exact calculations of permutation distributions for r independent samples. *Biometrical J*. 1985; 27: 431–43.

[409] Zumbo BD. Randomization test for coupled data. *Perception & Psychophysics*. 1996; 58: 471–78.

Index